IPTV
Opportunities

By Lawrence Harte

Althos Publishing
Fuquay-Varina, NC 27526 USA
Telephone: 1-800-227-9681
Fax: 1-919-557-2261
email: info@althos.com
web: www.Althos.com

Althos

All rights reserved. No part of this book may be reproduced or transmitted in any form or by any means, electronic or mechanical, including photocopying recording or by any information storage and retrieval system without written permission from the authors and publisher, except for the inclusion of brief quotations in a review.

Copyright © 2008 By Althos Publishing
First Printing

Printed and Bound by Lightning Source, TN.

> Every effort has been made to make this manual as complete and as accurate as possible. However, there may be mistakes both typographical and in content. Therefore, this text should be used only as a general guide and not as the ultimate source of IP Television information. Furthermore, this manual contains information on telecommunications accurate only up to the printing date. The purpose of this manual to educate. The authors and Althos Publishing shall have neither liability nor responsibility to any person or entity with respect to any loss or damage caused, or alleged to be caused, directly or indirectly by the information contained in this book.

International Standard Book Number: 1-932813-89-6

Acknowledgements

There were many people who provided me with information necessary to complete this book. They include Rick Sailor from Aminocom, Chris Wagner from Neulion, Roger McGarrahan from Pathfinder, Donald Cook from Falcon Communications, Taras Bugir from Harris, Ed Coughlin from Optibase.

Thanks to the many companies that have allowed interviews to share their knowledge and skills to help define the opportunities described in this book. They include Oracle, Microsoft, Ericsson, Calix, Harris, Siemens, Arris, Seachange, Digital Fountain, Orca, Espial, and many others. These companies took the time to provide information and explain important aspects of IPTV systems and services.

Special thanks to the many research companies who provide research data which include In-Stat, Point Topic, MRG, Infonetics, ITU, The Diffusion Group, Pricewaterhouse Coopers, and eMarketer. These companies gather and provide information that is necessary for business leaders and managers to make decisions on which products and services they will develop for the IPTV industry.

IPTV Business Opportunities

About the Author

Mr. Harte is the president of Althos, an expert information provider which researches, trains, and publishes on technology and business industries. He has over 29 years of technology analysis, development, implementation, and business management experience. Mr. Harte has worked for leading companies including Ericsson/General Electric, Audiovox/Toshiba and Westinghouse and has consulted for hundreds of other companies. Mr. Harte continually researches, analyzes, and tests new communication technologies, applications, and services. He has authored over 80 books on telecommunications technologies and business systems covering topics such as mobile telephone systems, data communications, voice over data networks, broadband, prepaid services, billing systems, sales, and Internet marketing. Mr. Harte holds many degrees and certificates including an Executive MBA from Wake Forest University (1995) and a BSET from the University of the State of New York, (1990).

IPTV Business Opportunities

Table of Contents

IPTV BUSINESS OPPORTUNITIES 1
   IPTV VALUE CHAIN 3
IPTV MARKETPLACE 4
   TELEVISION VIEWERS 4
   IPTV SUBSCRIBERS 5
   BROADBAND SUBSCRIBERS 7
     Digital Subscriber Line (DSL) 7
     Cable Modems 8
     Wireless Broadband 9
     Fiber Access Lines 11
   CABLE TELEVISION SUBSCRIBERS 12
   TELEVISION ADVERTISING 13
   TELEPHONE LINES 13
   IPTV EQUIPMENT MARKET UPDATE 14
   RATE PLANS 16
     IPTV Rate Plans 17
     Mobile TV Rate Plans 20
   MEDIA CONSUMPTION HABITS 21
   MOBILE VIDEO SERVICE REVENUE 22

ELECTRONIC COMMERCE (E-COMMERCE)	23
United States	*24*
Europe	*25*
Asia Pacific	*25*
INTERNET ADVERTISING	27
USER GENERATED CONTENT (UGC)	28

IPTV SERVICES ... 31

LINEAR TELEVISION CHANNELS	31
ON DEMAND PROGRAMMING	31
MEDIA PORTABILITY	31
IPTV ROAMING	33
INTERACTIVE APPLICATIONS	34
ADVERTISING	34
Advertising Metrics	*38*
Rules Based Advertising	*39*
Overlay Advertising	*39*
Contextual Advertising	*39*
Addressable Advertising	*40*
Personalized Advertising	*41*
User Profiling	*41*
Ad Telescoping	*41*
TELEVISION COMMERCE (T-COMMERCE)	42

THE KEY PLAYERS IN IPTV ... 49

TELEPHONE COMPANIES	49
CABLE SYSTEM OPERATORS	50
INTERNET SERVICE PROVIDERS (ISPs)	51
MOBILE CARRIERS	51
ELECTRIC UTILITIES	52
BROADCASTERS	53
WIRELESS BROADBAND	53
INTERNET TV	54

IPTV CONTENT – THE MILLION CHANNEL NEED 59
NETWORK PROGRAMMING 60
ORIGINAL PROGRAMMING 60
SPONSORED CONTENT 61
COMMUNITY CONTENT 61
INDEPENDENT CONTENT 62
Shared Content *62*
Wiki TV *62*
LOCAL PROGRAMMING 63
INTERNATIONAL PROGRAMMING 63
Regulatory Differences *64*
Program Access Controls *65*
Language Variations *65*
Time Offset *65*
Media Formats *66*
Program Guides *66*
Advertising *66*
COMPANY PROGRAMMING 67
PERSONAL PROGRAMMING 67
CONTENT PARTNERS 69
WHOLESALE ON-DEMAND 69

CONTENT MANAGEMENT 71
CONTENT LIFECYCLE 71
Long Tail Content *72*
Short Tail Content *72*
Flat Tail Content *72*
METADATA 73
CONTENT WORKFLOW 73

HOW IPTV AND INTERNET TELEVISION SYSTEMS WORK 75
DIGITIZATION - CONVERTING VIDEO SIGNALS AND AUDIO SIGNALS TO DIGITAL SIGNALS 75

- DIGITAL MEDIA COMPRESSION – GAINING EFFICIENCY 76
- SENDING PACKETS 78
  - *Packet Routing Methods* *78*
  - *Packet Losses and Effects on Television Quality* *79*
  - *Packet Buffering* *80*
- CONVERTING PACKETS TO TELEVISION SERVICE 82
  - *Gateways Connect the Internet to Standard Televisions* *82*
- MANAGING THE TELEVISION CONNECTIONS 83
  - *Switching (Connecting) Media Channels* *83*
- MULTIPLE IPTVs PER HOME 85
- TRANSMISSION 86
  - *Unicast* *86*
  - *Multicast* *87*
- CHANNEL SELECTION 89
  - *Program Guide* *89*
  - *Recommendation Engine* *91*
- ADDRESSABLE ADVERTISING 91
- VIDEO ON DEMAND (VOD) 92
  - *Download and Play* *93*
  - *Streaming* *95*
  - *Progressive Downloading* *97*
  - *Push Video on Demand (PVOD)* *97*
- HOME NETWORK MANAGEMENT 97
- SERVICE PROVISIONING 98
  - *Activation* *98*
- CONDITIONAL ACCESS SYSTEM (CAS) 99
- DIGITAL RIGHTS MANAGEMENT (DRM) 99

IPTV SYSTEMS 103

- VIEWING DEVICES 103
  - *Multimedia Computer* *104*
  - *IP Televisions* *105*
  - *3 Dimensional Displays (3D Displays)* *105*
- IP SET TOP BOXES (IP STB) 108
  - *Sensory Accessories* *108*

Table of Contents

  PREMISES DISTRIBUTION 108
    Wired LAN *109*
    Telephone Wiring *109*
    Coaxial Cable *110*
    Wireless LAN *111*
    Power Line Wiring *111*
  BROADBAND ACCESS NETWORK 112
    Digital Subscriber Line (DSL) *114*
    Cable Modem *116*
    Wireless Broadband *118*
    Power Line Carrier (PLC) *123*
  HEADEND 125
  CONTRIBUTION NETWORK 127

IPTV SYSTEM TYPES **131**
  ON NETWORK (ON-NET) 134
  OFF-NETWORK (OFF-NET) 134
  OVER THE TOP 134
  HYBRID IPTV SYSTEMS 135
  SYSTEM OPERATION OPTIONS 135
  NETWORK OPERATOR 135
  SHARED NETWORK 136
  VIRTUAL (HOSTED) NETWORK 137

PRIVATE IPTV SYSTEMS **139**
  PRIVATE IPTV BENEFITS 140
  PRIVATE IPTV EQUIPMENT 142
  PRIVATE TV CONTENT SOURCES 142

IPTV ECONOMICS **145**
  REVENUE 145
    Basic Services *146*
    Extended Services *146*
    Premium Services *146*
    Internet Access *146*

- *Voice Services* 146
- *Advertising* 146
- *Digital Services* 147
- *Television Commerce (T-Commerce)* 147
- *Other* 148
- COSTS 149
  - *Content Costs* 149
  - *Operations* 150
  - *General Administration* 150
  - *Marketing* 150
  - *Data Services Costs* 151
  - *Voice Service Costs* 152
- CAPITAL COSTS 153
  - *Headend* 154
  - *Middleware* 154
  - *Conditional Access and DRM System* 154
  - *Customer Premises Equipment (CPE)* 155
- OTHER COSTS 156
  - *Financing Cost* 156
  - *Subscriber Acquisition Cost (SAC)* 156
  - *Post Sales Support* 156
  - *Churn* 156
  - *Billing Systems* 157

IPTV CHALLENGES 159

- CONTENT DISTRIBUTION RIGHTS 159
- IPTV SYSTEM CAPACITY 161
- CONTENT PROTECTION 163
  - *Content Protection Options* 165
  - *Studio Endorsements* 166
- TELEVISION QUALITY 167
  - *Audio Quality* 167
  - *Video Quality* 169
- RELIABILITY 170
  - *Access Device Reliability* 170

Data Network Reliability *172*
Data Connection Reliability *172*
IPTV Server Reliability *174*
Feature Operation Reliability *175*
REGULATION 176
PRIVACY REQUIREMENTS 176
HDTV OVER IPTV 177
PATENT LICENSING 177
MEDIA PLAYER COMPATIBILITY 178
CHANNEL CHANGE TIME 179
IPTV SECURITY 180

IPTV SYSTEM INTEGRATION **183**
SYSTEMS INTEGRATORS 184
QUALIFYING AN IPTV SYSTEMS INTEGRATOR 185

APPENDIX I - ACRONYMS **189**

APPENDIX II - RESEARCH COMPANIES. **191**

INDEX. **195**

Chapter 1

IPTV Business Opportunities

IPTV business opportunities are the revenue producing services or products that people or companies can offer into the Internet protocol television (IPTV) industry. IPTV business opportunities range from getting a job in a high-growth IPTV company to becoming a global television broadcast company.

Some of the opportunities for the IPTV industry are not new concepts. They are simply new ways to provide similar types of services. However, IPTV systems do have capabilities that traditional television systems cannot provide in their existing forms. For example, IPTV systems can provide over 1 million television channels. To provide 1 million channels on satellite, television broadcast or cable TV systems, significant changes would need to be made.

IPTV is a process of sending television signals over the Internet or other types of data networks. If the television signal is in analog form (standard TV or HDTV), the video and audio signals are first converted to a digital form. Packet routing information is then added to the digital video and voice signals so they can be routed through the Internet or data network.

To get IPTV, you need these key parts:
- Viewing devices or adapters
- Broadband access providers
- IPTV service providers
- Media content providers

The viewing devices or adapters convert digital television signals into a form that can be controlled and viewed by users. Broadband access providers supply the high-speed data connection that can transfer the digital video television signals. Service providers identify and control the connections between the viewing devices and the content providers (media sources). Media content providers create information that people want to view or obtain.

Figure 1.1 shows a sample IPTV system. This diagram shows the IPTV system gathers content from a variety of sources including network feeds, stored media, communication links and live studio sources. The headend converts the media sources into a form that can be managed and distributed. The asset management system stores, moves and sends out (playout) the media at scheduled times. The distribution system simultaneously transfers multiple channels to users who are connected to the IPTV system.

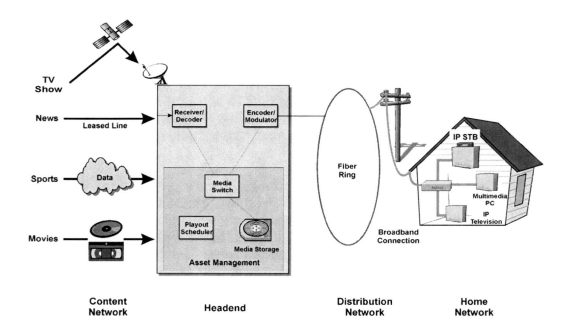

Figure 1.1, IPTV System

Users view IPTV programming on analog televisions that are converted by adapter box (IP set top box), on multimedia computers or on IP televisions (data only televisions).

IPTV Value Chain

A value chain is an operational model that describes the core functions that are required to deliver products or services to the end customer. The blocks in a typical IPTV value chain include content producer, content aggregator, content distributor and content consumer.

Content producers are companies or developers of media content. Content producers may directly provide distributors or network providers with access to content. Content produces ad value by converting ideas or creative components into media in tangible formats.

A content aggregator obtains the rights from multiple content providers to resell and distribute through other communication channels. A content aggregator typically receives and reformats media content, stores or forwards the media content, controls and/or encodes the media for security purposes, accounts for the delivery of media and distributes the media to the systems that sell and provide the media to customers.

A content distributor (such as a broadcaster) receives content from one or more companies (typically content aggregators), stores and manages the content and transfers content to one or more consumers.

Content consumers receive and use media or data. The permitted uses of media or data by content consumers may be restricted in time, format or other usage criteria.

Figure 1.2 shows a sample IPTV value chain. This value chain starts with a content producer (movies or shows), media (games) or services that they desire to sell to content viewers. The content producer provides the content to content aggregators who gather, process and provides the programs when required to carriers. Carriers (broadcasters) order, manage and deliver the programs to customers through their distribution systems.

Figure 1.2, IPTV Value Chain

IPTV Marketplace

The IPTV marketplace is a result of the combination of television, broadband data connections, and changing consumer media consumption habits.

Television Viewers

There are approximately 1.2 billion TV households [1] and more than 1.7 billion television sets worldwide [2]. According to A.C. Neilson and OECD, the average television viewing time continues to increase for most countries with the US having the highest daily viewing time of 8 hours and 14 minutes per TV household [3]. In Europe, OECD reports that the average viewing time ranges from 2.75 to 5 hours daily viewing time per TV household [4].

According to Solutions Research Group (SRG), television viewers have been changing their viewing habits. Nearly 80 million Americans (43%) have begun watching some of their favorite television shows through the Internet. Approximately 25% of prime time viewing was time shifted using a media storage device such as a DVR. When these consumers watched stored programs, 65% said they "always" skip through commercials [5].

IPTV Subscribers

According to Multimedia Research Group (MRG), the number of IPTV subscribers will grow from 14.3 million in 2007 to 63.6 million in 2011, a compound annual growth rate of 45 percent.

In 2007, Europe was the biggest market for IPTV. France easily leading the growth spurt through IPTV operators Free, Orange France Telecom and Neuf Cegetel. Also driving the market's successful growth in the first half of 2007 was the fast growth in other parts of Europe, especially Belgium, Spain, Italy and Eastern Europe. In Asia, growth has been especially high in China, Japan and Hong Kong. In North America, growth has been robust in Canada and the United States, especially from Verizon and the Independent Operating Companies (IOCs).

Figure 1.3 shows the global IPTV subscriber forecast from 2007 to 2011 will grow from 14.3 million in 2007 to 63.6 million in 2011, a compound annual growth rate of 45 percent. This chart shows that Europe remains the leading region and its subscriber numbers have increased versus the October 2006 forecast. Asia's prospects have continued to improve as tests in China and India have blossomed into successful deployments. MRG forecasts that Asian and European subscriber numbers will steadily grow closer from 2007 to 2011.

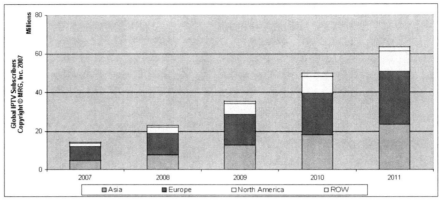

Figure 1.3 , IPTV World Subscriber Forecast
Source: IPTV Global Forecast Report—April 2007, www.mrgco.com

The service revenue projections by MRG show dramatic increases in Europe, North America and Asia 2008-2011. The forecast also predicts that average revenue per user (ARPU) will grow due to addition of services by existing subscribers.

Figure 1.4 shows the global IPTV service revenue forecast from 2007 to 2011. This chart shows that IPTV service revenue will increase from $3.6 billion in 2007 to $20.3 billion in 2011, a compound annual growth rate of 71 percent. The service revenue projections by MRG show dramatic increases in average revenue per user (ARPU) due to added services.

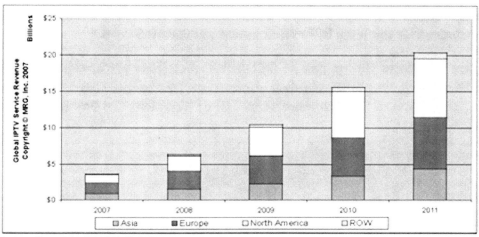

Figure 1.4 , IPTV World Revenue Forecast
Source: IPTV Global Forecast Report—April 2007, www.mrgco.com

Broadband Subscribers

Broadband subscribers are people who can receive data services that have transmission rates of 1 Mbps or higher. Of the 1.32 billion Internet users worldwide in 2007 [6], there were 328.8 million broadband subscribers worldwide with a positive growth rate in each country [7]. Broadband subscription growth is strong but slowing in all continents during 2008. The key types of broadband access lines for IPTV include DSL, cable modem, fiber and wireless.

Digital Subscriber Line (DSL)

DSL continues to lead the broadband industry with 66% broadband market keeping well ahead of cable modem, fiberoptic and wireless broadband connections. The percentage of subscribers who choose DSL when it is available (take up rate) varies based on region. In some countries, DSL accounts for almost 100% of market share while in other countries (such as North America) cable modems have the highest penetration rate.

Figure 1.5 shows the growth of DSL subscriber lines worldwide from 2003 to 3rd Quarter 2007. This graph shows that DSL subscribers have grown from 28 million in 2003 to more than 210 million near the end of 2007.

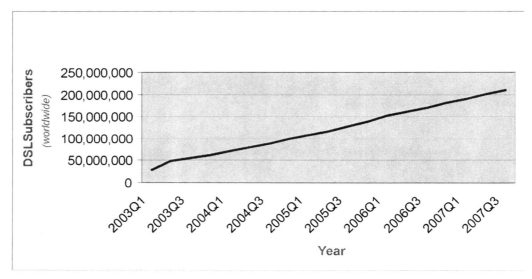

Figure 1.5, Worldwide DSL Lines 2003 to 2007
Source: Point Topic

Cable Modems

The cable modem market was 71 million households in 2007. Cable modem services revenues increased from $22 billion in 2005 to $26 billion in 2006. The broadband cable services market has aggressive competition from DSL (outside the USA) and competition is likely to increase with wireless broadband (e.g. WiMAX) and high speed mobile communication (e.g. 3G cellular).

Figure 1.6 shows that the worldwide cable modem market has steadily grown to over 71 million customers by 2007. This diagram shows the while the growth of new cable modem subscribers is high, the growth rate has decreased from 28% in 2003 to 18% in 2007 [8].

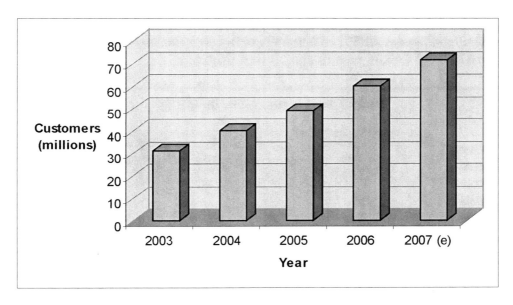

Figure 1.6, Worldwide Cable Modem Market 2003 to 2007
Source: In-Stat

Wireless Broadband

There are several types of wireless systems that can provide high-speed broadband data services. The leading wide-area wireless broadband system in 2008 was WiMAX.

According to market intelligence company Maravedis, the growth of WiMAX subscribers increased more than 85% between 1Q 2006 and 1Q 2007. The number of broadband wireless subscribers worldwide grew to 950,000 with 300,000 of these subscribers using industry standard WiMAX devices.

Maravedis research showed the WiMAX systems in the Asia provided the highest data transmission speed (typically 1.6 to 2.1 Mbps) but had the lowest average revenue per user ($30.45). The ARPU for business ($145.54) was much higher than residential ($40.76) and the mix of residential to business customers was 58% residential to 42% business.

Figure 1.7 shows the average revenue per user (ARPU) for residential broadband wireless access by geographic region in 2007. This diagram shows that the lowest ARPU of $30.45 was in the Asia Pacific region and the highest ARPU of $46.56 was in the United States.

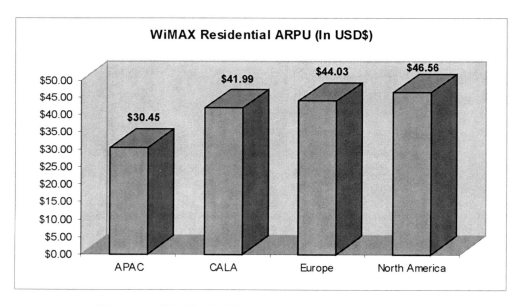

Figure 1.7, Worldwide Wireless Broadband ARPU in 2007
Source: Maravedis

The top countries for WiMAX deployments in the beginning of 2007 were the United States, Spain and Australia with a combined subscriber base of approximately ½ million subscribers. The growth of wireless broadband in nations that have developed alternative wireless broadband solutions (such as Korea) have been much slower.

Figure 1.8 shows the number of wireless broadband subscribers by geographic region in 2007. This diagram shows that the lowest number of wireless broadband subscribers (153k) was in central and Latin America (CALA) with the largest number of subscribers (359k) was in the United States.

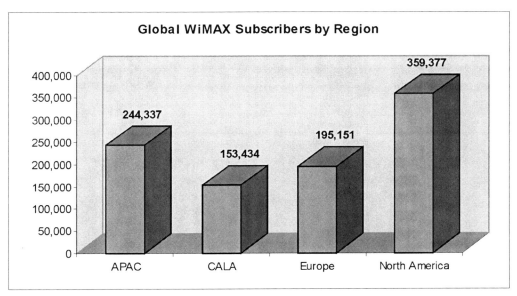

Figure 1.8, Worldwide Wireless Broadband Subscribers in 2007
Source: Maravedis

Fiber Access Lines

The number of fiber access line connections was approximately 20 million at the end of 2007 and it is expected that the number of fiber connections will reach more approximately 90 million by 2012 (about 5% of all homes worldwide) [9].

Fiber access is important to IPTV for increasing bandwidth needs such as increases in the resolution of television (HDTV), advanced media applications such as virtual reality and 3D television (60 Mbps+ per television).

Cable Television Subscribers

At the end of 2007, there were more than 408 million are cable TV households. [10]. The growth of cable television subscribers from 2006 to 2007 was 10%. The largest countries with cable television subscribers are China and the USA.

Figure 1.9 shows that the worldwide cable television market has steadily grown to over 355 million customers by 2007. This diagram shows the growth rate of new cable television subscribers has been 2% to 3% each year.

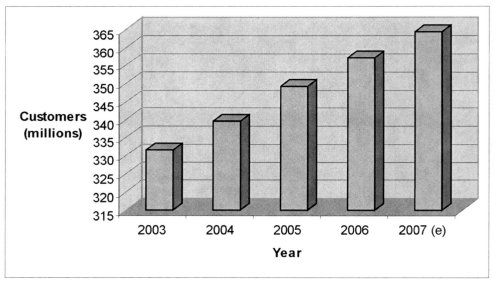

Figure 1.9, Cable Worldwide Cable Television Market Growth 2003 to 2007
Source: Instat

Television Advertising

Television advertising is the sending of promotional messages or media content to one or more potential television program viewers. The global market for all advertising is expected to increase from $449 billion in 2007 to $479 billion in 2008 [11]. Advertising spending is split between television (approximately 1/3rd), radio, magazines, newspapers and the Internet. The United States is the leader in television advertising revenue with $48.4 billion spent on TV ads in 2006 ($161 per person each year) [12].

Advertising spending tends to increase with the global economy. However, advertising spending is shifting from television advertising to other forms of advertising such as Internet advertising and mobile advertising.

IPTV systems can use addressable advertising, which allows different ads (via separate streams or on demand ads) to be sent to different TV sets during the same advertising time slot. Using addressable advertising, ads can be tailored according to individual demographics; location; interests; viewing habits; time of day; language and a raft of other factors. This can increase the effectiveness of television ad campaigns resulting in much higher ad rates.

Telephone Lines

Telephone lines are access connections from users to communication service providers. Fixed service telephone lines can use copper, coax, fiber or wireless links as the connection method. In 2007, there were approximately 1.3 billion fixed telephone lines worldwide [13].

In the 1990s and early 2000s, a dramatic shift occurred in the telecom industry. Customers were transitioning from fixed wired access lines to mobile communication lines. The number of fixed lines is still important to IPTV as these lines can be converted to broadband data connections (DSL) relatively easily.

Figure 1.10 shows the number of fixed telephone lines worldwide. This graph shows that the number of fixed telephone lines increased from 738 million lines in 1996 to 1.27 billion lines in 2006.

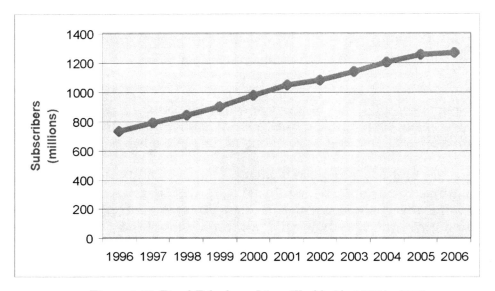

Figure 1.10, Fixed Telephone Lines Worldwide 1996 to 2006
Source: International Telecommunications Union (www.ITU.int)

IPTV Equipment Market Update

IPTV equipment is the hardware devices and assemblies that are used to receive, process and display television signals in IP video format. IPTV equipment includes head end equipment, media servers, distribution equipment and set top boxes.

Research company Infonetics predictions show that sales of IPTV equipment are forecast to skyrocket from over $1 billion in 2006 to near $6 billion in 2010.

Chapter 1

A key IPTV equipment market growth indicator was the increase in IP STB revenue of 35% in 3rd Quarter 2007. The sales of IP STBs are expected to be strong as more models are available with MPEG-4 HD. The number of IP set top boxes (STBs) sold worldwide was led by Motorola for pure IP STB and hybrid Cable/IPTV and by Dasan for hybrid DTB/IP STB [14].

In 3Q07, equipment sales were divided 51% from Asia Pacific, 33% from EMEA, 16% from North America, and 1% from CALA.

Figure 1.11 shows that IPTV equipment sales in 2006 was more than $1 billion and that Infonetics projects the sales of IPTV equipment will increase to almost $6B by 2010.

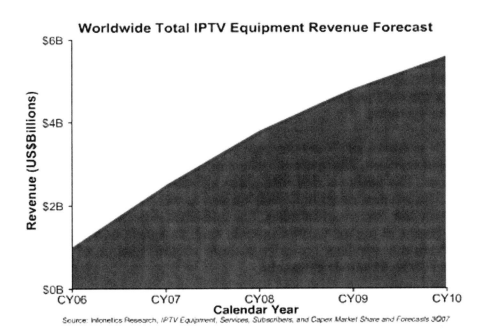

Figure 1.11, Worldwide IPTV Equipment Revenue 2006 to 2010
Source: Infonetics (www.Infonetics.com)

15

Rate Plans

A rate plan is the structure of service fees that a user will pay to use services. Rate plans are typically divided into monthly fees and usage fees.

IPTV system operators initially offered services and rate plans that directly competed with cable and satellite television systems. IPTV rate plans are typically composed of competitive monthly access fees ($30 to $50 per month) that provide unlimited access to 30-50 basic television channels (news, weather, sports and entertainment) and added cost options for premium movie channels. IPTV systems are beginning to differ is on the number of available music video and television channels. There can be several thousand more channels, as compared to the 10 to 50 video on demand channels, offered by traditional cable television systems.

As the IPTV industry has matured, IPTV system operators have started to take advantage of being able to provide access to virtually an unlimited number of television channels. This means IPTV service providers can offer thousands or tens of thousands of channels and programs compared to existing television systems.

Like cable TV systems, IPTV service rate plans typically include at multiple levels (tiers) of services. These often include basic service, mid-level and premium services. The basic service rate plans typically includes several local and regional television programs such as news, weather, community and other relatively low cost programming. The mid-level service plans often add one or two groups of higher value channels such as sports and some movie channels. Premium services typically include several groups of premium channels and 30 to 45 music channels.

Some of the key rate plan differences include the individual pricing of channels, enhanced navigation options and an increase in the number of international channels.

IPTV systems such as NowTV in Hong Kong offer some or all of their channels by individually charging for each channel. The user can simply select which channels they want and avoid paying for channels that do not interest them. IPTV systems may provide enhanced channel guides, which allowed the viewers to search and find specific programs and program types. Many of the IPTV systems in the early 2000s offered viewers with the option to obtain a variety of international channels. While these international channels may not have a high take-up rate, they offer significant value to foreign nationals who are living in another country or who have relatives or friends who are located in other countries.

Another feature for IPTV service providers was their focus on using online customer service applications as opposed to encouraging customers to call a service center for assistance with initiating a service order. Some IPTV service providers charge extra fees for receiving printed billing statements as opposed to receiving them online and automatically paying them by credit card.

IPTV Rate Plans

IPTV and cable TV service rate plans can be hard to compare. Many of the service rate plans were dependent on the specific location for the customer. It appears that channel groups change between regions and could even rapidly change within a specific area.

Figure 1.12 shows sample IPTV and cable TV rates throughout Asia, Europe and the USA. The IPTV and cable TV systems selected for this table were located in the same city. This table shows that the television service rates are often divided into basic, mid-level and premium groups. This table shows that while some IPTV rates are less than the cable (e.g. basic rates), others rates can be higher due to the inclusion (and high cost) of premium channels.

Company	System Type	Basic	Mid-Level	Premium
Allendale Communications Allendale, MI USA	IPTV	$12.95 (20 chan)	$42.95 (76 chan)	$52.95 (140 chan+45 music)
Now TV Hong Kong (exchange rate used =.13)[1]	IPTV	$51.22 (35 chan+)	$110.89 (75 chan+)	$128.83 (75 chan+ special pack)
Home Choice- UK (exchange rate used = 1.28)	IPTV	$23.03 (35 chan)	$35.82 (60 chan)	$40.63 (85 chan)
Charter Communications, Allendale, MI-USA	Cable	$54.99 (76 chan + 1 groups)	$65.99 (76 chan + 2 groups premium)	$70.99 (76 chan + 1 group + all premium)
Cable TV - Hong Kong Hong Kong[2] (exchange rate used =.13)	Cable	$40.04 (63 chan)	$49.14 (70 chan)	$59.94 (76 chan)
NTL Cable-UK (exchange rate used = 1.89)	Cable	$7.04 (top 10 UK chan+)	$14.72 (100 chan)	$24.96 (160 chan)

Notes: 1. NowTV rates calculated by adding individual fees for each channel.
2. Cable TV Hong Kong rates calculated by adding 1 program group, other program groups were available

Figure 1.12, IPTV Service Rate Comparison

Some of the fees associated with IPTV service include an installation fee, equipment rental fees, deposits, monthly service fees and pay per view fees.

The installation fee typically ranges from $30 to $150. There are various options for waiving (removing) the installation fee ranging from having another service, special promotions or some other event that encourages the customer to take immediate action.

Equipment rental fees for the set top box (commonly called a converter box) ranged from $5 to $30 per month. There was typically no rental fee charged for the broadband modem.

Deposits range from around $0 to $300. The amount of the deposit is affected by the length of the service contract (monthly, 1 year+).

Monthly service fees range from $13 per month (20 channels) to over $128 per month (200 channels+). Several IPTV systems offered the option for prepaid yearly fees, which are at a discount of 15% to 20% off the monthly rates.

Pay per view fees range from around $1 to $5 per view. Special viewing events could be considerably higher than standard PPV charges.

Figure 1.13 shows a sample IPTV service rate plan. This table shows that IPTV systems typically charge a setup fee of $30 to $150 and have equipment rental fees ranging from $5 to $30 per month. Some IPTV service providers do not require (waive) deposits and some do require deposits to ensure equipment is returned and not abused. Fees for IPTV monthly services ranged from $13 per month to more than $128 per month. Pay per view fees (PPV) typically cost $1 to $5 per view.

Fee	Cost
Installation	$30-$150
Equipment Rental	$5 to $30
Deposits	$0 to $300
Monthly Service Fees	$13 to $128
Pay per View (typical)	$1 to $5

Figure 1.13, IPTV Service Rate Samples

There is likely to be a dramatic increase in the number of available television channels and pay per view base for IPTV systems over the next few years as IPTV service providers increase their competition with cable, broadcast and satellite TV companies. These channels are likely to include live international channels, regional content and specialty programming.

Mobile TV Rate Plans

The initial service rate plans for mobile video systems use a combination of basic access service rates for a limited number of popular TV channels (5 to 15 typical) with a few key exceptions. Mobile video system operators are using some unique promotions and options to attract and keep new customers.

Monthly service fees range from $4 to $25 per month. For growing systems (such as the TU Media in Korea), reductions in service rates and increases in program channels have increased the number of viewers to over 1.2 million in 2007 (according to BusinessWeek). This is approximately 2.5% market penetration of the 40+ million mobile users in Korea. Some systems required the combination of other services, such as mobile data (GPRS or EVDO), to obtain access to mobile TV services. While many systems offered free mobile TV trial periods, it appears that no companies offered mobile TV for free (advertising paid) in 2007.

Some of the common popular categories for mobile TV include news, sports, music and entertainment. Most mobile TV packages offer a mix of video and radio channels. Because many of the mobile TV systems use one-way broadcast systems (e.g. DVB-H), on-demand programs may not be available or are available on a limited basis. For mobile TV systems that do have two-way capability (such as 3G Mobile), video clips may only be available as part of an upgraded premium package.

Like traditional TV systems, mobile video service rate plans typically have levels (tiers) of services. These include basic service, mid-level and premium services. The basic service rate plans typically include 5 to 10 local and regional popular television programs such as news, weather, community and entertainment programming (such as soap operas and cartoons). The mid-level service plans often add one or two groups of higher value channels such as sports and some movie channels. Premium services typically include several groups of premium channels and potentially on-demand channels.

Figure 1.14 shows a selective sample mobile TV rates throughout Asia, Europe and the USA. This table shows that most mobile TV systems offer a very limited number of channels (below 10) and their service rates are approximately $1 to $3 per channel.

System	Number of Channels	Rate
Qualcomm FLO TV (USA)	8	$15
Verizon (USA)	9	$15-$25
TU Media (Korea)	16	$12
T-System (Germany)	8	$5-$15
VTC Mobile (Vietnam)	8	$6 (90,000 VND)
Debitel (Germany)	4	$13.5 (9.95 EU)
Orange (UK)	28	$ 20 (10 GBP)

Figure 1.14, Mobile TV Service and Rate Comparison

A dramatic increase in the number of available television channels and pay per view base for mobile video systems is likely to occur over the next few years as mobile TV service providers increase their competition with mobile telephone companies. These channels are likely to include live international channels, specialty programming and shared content.

Media Consumption Habits

Media consumption is amount of information that a person or device receives and processes over a period of time. Media consumption is primarily split between television, radio, Internet, magazines and newspapers.

Television viewing times continue to grow or remain stable. Internet consumption is transitioning from communication services (such as email) to content services (watching online television programs). Consumers are still spending relatively brief periods of time interacting with their computer as compared to longer viewing periods for television. Consumer media consumption is also shifting from watching or listening to scheduled programming to on demand programming.

Mobile Video Service Revenue

According to Infonetics, mobile video service revenue in 2006 was $200 million. There are a variety of options to receive mobile video including satellite, broadcast systems (e.g. DVB-H), mobile multicasting and mobile streaming. Infonetics predicts that the sales of mobile phone with video capabilities was $58 billion in 2006 and this will increase to $125 billion by 2010.

Infonetics predicts that there will be 46 million (11.7 million from DVB-H and 6 million MediaFLO) mobile video subscribers by 2010 with mobile video service revenues increasing to more than $12 billion. This is a compound annual growth rate of 135% over 5 years. Infonetics expects that governments will provide incentives for mobile video system operators as the deactivation of analog television systems occurs in various countries.

Figure 1.15 shows the Infonetics projection that mobile video service revenues will increase from $200 million in 2006 to over $12 billion in 2010.

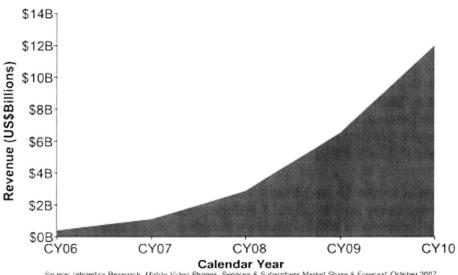

Figure 1.15, Worldwide Demand for Mobile Video Services
Source: Infonetics (www.Infonetics.com)

Electronic Commerce (e-commerce)

Electronic commerce (e-commerce) is a shopping medium that uses electronic networks (such as the Internet or telecommunications) to present products and process orders. E-commerce growth will continue to be high due to increasing amounts purchased by experienced online shoppers and by the addition of new online shoppers.

E-commerce sales values are important for IPTV systems as it is likely that television commerce (t-commerce) and mobile commerce (m-commerce) will become a significant part of e-commerce. If IPTV systems offer direct billing to IPTV subscriber accounts, e-commerce offers the highest potential for increasing average revenue per user (ARPU) for IPTV subscriptions.

According to Nielsen Company, by 2008, more than 875 million consumers (85 percent of the worlds Internet population) have used the Internet to make a purchase [15].

United States

According to the eMarketer Research, e-commerce in the United States during 2006 was $108.7 billing and it expects that e-commerce revenue will increase at a CAGR of 17% to $243.5 billion in 2011 [16]. According to the U.S. Census Bureau, the percentage of retail sales in the United States that comes from e-commerce as compared to total retail sales has increased from 1.8% in 2004 [17] to 3.4% in 2007 [18].

E-commerce is in the United States has matured and approximately 2/3rd of all Internet users in the US (117 million people) are expected to make e-commerce purchases in 2008 with the amount of purchases over the year of $1,123 [19].

Figure 1.16 shows the revenues generated by e-commerce in the United States region will increase from $108.7 billion in 2006 to $243.5 billion in 2011.

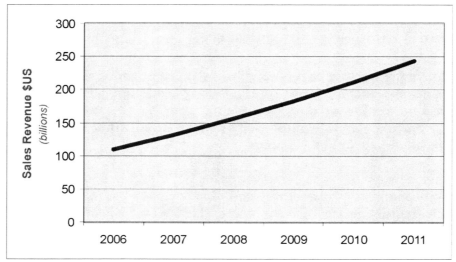

Figure 1.16, E-Commerce Revenue for US 2006 to 2011
Source: eMarketer (www.eMarketer.com)

Europe

In 2007 e-commerce sales in Europe was $175 billion and eMarketer Research estimates that this will reach $407 billion by 2011 (an annual growth rate of 25%). The combination of UK, Germany and France currently are 72% of the total e-commerce sales in Europe. Eastern European countries are a small portion of e-commerce but their share of e-commerce should rapidly grow.

Figure 1.17 shows the revenues generated by e-commerce for the UK, Germany and France in 2006. This table shows that the e-commerce sales for the UK were $55.6 billion, Germany was $27.1 billion and France was $12.5 billion. The expected compound annual growth rate for each country from 2006 through 2011 is expected to be over 20%. The UK had the highest average annual spending for Internet shoppers at $2,241 per online shopper.

Country	Sales (billions $US)	Expected CAGR 2006-2011	Online Buyers (millions)	Average Annual Spending
UK	$55.6	22.7%	24.8	$2,241
Germany	$27.1	24.1%	27.2	$996
France	$12.5	27.4%	14.5	$860

Figure 1.17, European e-Commerce for UK, Germany and France in 2006
Source: eMarketer (www.eMarketer.com)

Asia Pacific

According to eMarketer Research, e-commerce during 2007 in the Asia Pacific region was $73.3 billion. This was a 24% increase from the $59.1 billion in 2006. Japan was the largest market with 63% of e-commerce sales followed by Korea. It is expected that China and India will become the e-commerce leaders as the challenges of shipping and online financial transaction infrastructure and consumer confidence in e-commerce vendors increase.

Figure 1.18 shows the revenues generated by e-commerce in the Asia Pacific region will increase from $59.1 billion in 2006 to more than $168.7 billion in 2011.

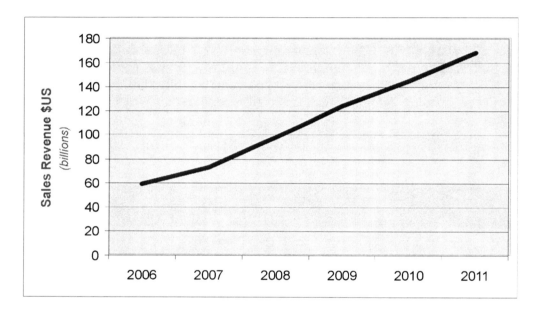

Figure 1.18, E-Commerce Revenue for Asia Pacific 2006 to 2011
Source: eMarketer

Internet Advertising

Internet advertising is the communication of messages or media content to one or more potential customers through the use of the Internet. According to the Internet advertising bureau (IAB), Internet advertising exceeded $16.8 billion in 2006 and is growing at more than 25% per year [20].

According to the IAB, the key portion of Internet advertising spending is on search engine advertising [21]. This is transitioning advertising from push broadcasting to pull consumer. Internet advertising is likely to be merged into IPTV systems to provide high-value individual targeted advertising options.

Figure 1.19 shows how Internet advertising revenue has increased from $4.6 billion in 1999 to more than $16.8 billion in 2006.

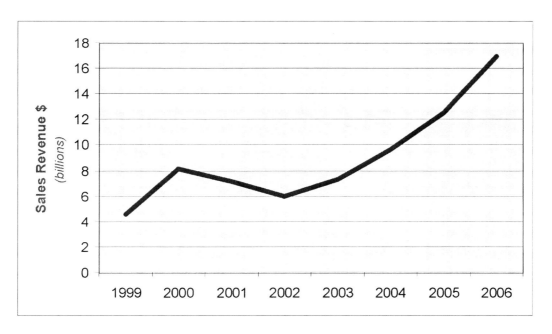

Figure 1.19, Internet Advertising Revenue 1999 to 2006
Source: PwC/IAB Internet Advertising Revenue Report (www.iab.net)

User Generated Content (UGC)

User generated content is information that is produced by users of similar content. There were more than 22.4 billion user generated videos viewed in 2007, which was an increase of 70% from the 13.2 billion videos viewed in 2006. According to Accustream iMedia research, the average number of views for each user generated video was 10,695 [22].

UGC is commonly social videos that are created by amateur, semi-professional and professional people or companies for the purpose of sharing them with other viewers. Successful UGC videos have contain various themes including controversial subjects, shock value (unexpected content), being risqué or being humorous. Social videos may be distributed through social networks or directly between viewers (viral distribution).

Figure 1.20 shows the number of social (user generated) videos and the predicted number of viewings of social videos in 2008. This table shows that the number of views increased by 70% from 2006 to 2007 reaching over 22 billion video views.

Year	Views (billions)
2005	3.3
2006	13.1
2007	22.4
2008 (Est)	34.0

Figure 1.20, User Generated Videos
Source: *AccuStream iMedia Research (www.accustreamresearch.com)*

UGC can be a low cost source of content for IPTV systems. Trends are for higher quality and longer social videos. Examples of social video portals include YouTube, Yahoo Video and Google Video. Social videos can generate revenue from advertising sponsorship in the video server portal (banner ads) and 5 to 15 second ads that are played before the video is provided.

Figure 1.21 shows the average length of user generated content on some of the leading shared online video web sites for 2007. This table shows that some of the web sites specialize in longer video files.

UGC Web Site	Average Length (minutes)
Crackle.com	5.13
Revver.com	2.53
MySpace.com	3.00
Veoh.com	17.10
Yahoo Video	4.05
YouTube	3.21

Figure 1.21, User Generated Content Video Length
Source: *AccuStream iMedia Research (www.accustreamresearch.com)*

References:

1. Worldwide Cable TV Households Continue to Grow as Cable Operators Expand Offerings", In-Stat Press Release, www.instat.com, 8 Nov 2006.
2. "Microsoft May be a TV Star Yet", Business Week, www.BusinessWeek.com, 7 Feb 2005.
3. "Digital Video Recorders Grow in Popularity", The Nielsen Company, www.nielsenmedia.com, 17 Oct 2007.
4. "OECD Communications Outlook 2007", Organization for Economic Co-Operation and Development, www.OECD.org, 2007
5. "Prime Time is Anytime", Solutions Research Group, www.srgnet.com, 12 Feb 2008.
6. "The Internet Big Picture, World Internet Users and Population Stats", www.internetworldstats.com
7. "Global broadband growth is slowing down", Point-Topic, www.Point-Topic.com, 12 Jan 2008.
8. In-stat Cable Modem, xxx, www.in-stat.com
9. "FTTH Worldwide Technology Update & Market Forecast", Heavy Reading, www.HeavyReading.com
10. "The Worldwide Market for Cable Television Services," www.Instat.com, In-Stat, December 2007
11. "US Ad Spend to Grow 3.7% in '08, Up from 2.8" in '07; Global to grow 7%", Group M press release, www.GroupM.com

12. "Research Central, Ad Revenue Track, 2006, FULL YEAR 2006 SUMMARY", television bureau of advertising, www.tvb.org
13. "Key Global Indicators for the World Telecommunication Service Sector", International Telecommunications Union, www.itu.int.
14. Infonetics IPTV Equipment Forecast xxx.
15. "Over 875 Million Consumers have Shopped Online – The number of Internet Shoppers Up 40% in Two Years", The Nielsen Company, New York, www.nielsen.com, January 28, 2008.
16. "US Retail E-Commerce Sales Maturing", eMarketer Research, www.eMarketer.com, 21 May 2007.
17 "Quarterly Retail E-Commerce Sales 2nd Quarter 2005", U.S. Census Bureau News, www.census.gov, 19 Aug 2005.
18. "Quarterly Retail E-Commerce Sales 3rd Quarter 2007", U.S. Census Bureau News, www.census.gov, 19 Nov 2007.
19. "US Retail E-Commerce Sales Maturing", eMarketer Research, www.eMarketer.com, 21 May 2007.
20. "Internet Advertising Revenues in Q3 '07 Surpass $5.2 Billion, Setting New High," www.IAB.net, November 12, 2007
21. "IAB Internet Advertising Report 2006", Internet Advertising Bureau, May, 2007, www.IAB.net.
22. "User Generated Video 2005 – 2008: Mania Meets Mainstream," Accustream iMedia Research, www.Accustreamresearch.com.

Chapter 2

IPTV Services

IPTV services include providing program content, interactive services, advertising and television commerce.

Linear Television Channels

Linear television is the providing of television programs in a time sequence (scheduled programming). Linear television programs may be live or stored content. Linear programs may be paid for by advertising (ad sponsored) and/or based on a subscription service (premium programs).

On Demand Programming

On demand programming is providing or making available programs that users can interactively request and receive. On demand programs may be paid for by a subscription plan (such as an unlimited viewing of a premium program channel) or on a pay per view or pay per use basis.

Media Portability

Media portability is the ability to transfer media from one device or storage area to another device or storage area. IPTV service providers may allow or charge for providing the ability to transfer media programs between users or devices. An example of media portability is the ability to download a song from iTunes to a personal computer (PC) and have the ability to transfer the music from the PC to an iPod.

To provide media portability services, the IPTV service provider needs to manage connection capability and obtain media portability rights. Portability rights are the permissions granted from an owner or distributor of content to transfer the content to other devices (such as from a set top box to a portable video player) and other formats (such as low bit rate versions).

Figure 2.1 shows how a media portability management system may be used to allow users to transfer media from one device to another. This example shows how a person has downloaded a TV program to their set top box. The media portability management system has setup authorization for the

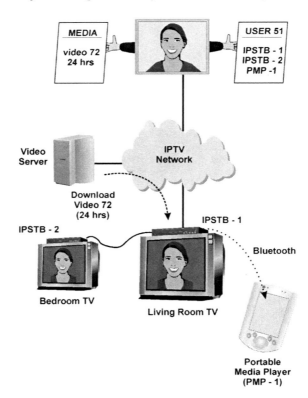

Figure 2.1, Media Portability Management

media to be transferred throughout the user's personal domain for a period of 24 hours. This allows the person to access and view the program through other devices in the home and to transfer it to their personal media player.

IPTV Roaming

IPTV roaming (also known as place shifting) is the capability for an IPTV subscriber to access the content (e.g. television programs) offered by their IPTV service provider through other communication systems (such as broadband data connections in hotel rooms).

IPTV systems are likely to expand their networks into other countries through the use of broadband Internet access systems. IPTV operators can leverage their relationship with content suppliers and use their asset management systems to offer television services in other countries without having to own or even operate the broadband access networks. IPTV roaming may be performed via service providers or by user owned and operated television gateways (such as Slingbox™).

User owned TV gateways (such as a Slingbox) allow individuals to extend the television services with no additional monthly service fees. This enables consumers to watch their cable, satellite or digital video recorder (DVR) programming from wherever they are. All they need is an Internet-connected laptop or desktop PC, a TV gateway and a broadband connection in their home. The TV gateway redirects or place shifts a live TV stream through a broadband Internet to a PC located anywhere in the world.

From a technical perspective, the TV gateway receives video signals from a cable line or television accessory, digitizes and compresses the analog signals and retransmits the digital signals so they can be viewed on a media viewing device. In essence: "TV video in" and "IP stream media out".

IPTV systems that provide roaming services need to allow for the user to operate even in very, very low bit rates and to help the user understand what is going on. The user may be provided with status information such as available bandwidth. Access providers (such as hotel data systems) may intentionally limit the data transmission bandwidth available to the user. This means that the network conditions may continually vary. The systems will need to dynamically adjust the video compression ratios to match the available network bandwidth.

Interactive Applications

Interactive applications are software programs that are designed to perform operations using commands or information from users (such as a user selecting an option via a keyboard). Popular IPTV interactive applications include gaming, shopping and expandable advertising. IPTV service providers may charge service fees (such as additional subscription fees) for users to have access to interactive applications or they may provide free interactive applications as a marketing tool to help differentiate their IPTV services from traditional television broadcast applications.

Advertising

One of the most complicated new areas for IPTV billing is likely to be the management of advertising services. Advertising management is the process of creating, presenting, managing, purchasing and reporting of advertising programs. Because advertising services on IPTV systems can range from broadcast advertising (on to all people in a geographic area) to customized addressable advertising (custom ads for specific viewers), advertising management can be a complex and very profitable process. For example, at the beginning of 2006, the highest cost of advertising for broadcast television was approximately 2 cents per impression (the Superbowl) compared to the per viewer web marketing cost of 54 cents for Google ad clicks.

An advertising program typically begins by setting up advertising campaigns. Advertising campaigns define the marketing activities such as the specific advertising messages that will be sent to customers who are classified into certain categories (target market segments) about products, services and options offered by a company.

Chapter 2

IPTV advertising messages may be in the form of interstitial, mixed media or interactive media. An interstitial ad is an advertising message that is inserted "in between" program segments. Interstitial ads can also be pop-ups (when selecting a new channel) and pop-downs (when exiting a selected program). Mixed media advertising is the combining of advertising media along with other video and text graphics on a television or video monitor. Interactive advertising is the process of allowing a user to select or interact with an advertising message.

Figure 2.2 shows how IPTV advertising messages can be in the form of interstitial broadcast messages, mixed media messages or interactive ads. In example A, a network operator provides a program with advertising messages already inserted (interstitial) into the program. Example B shows how an advertising message may be overlapped or merged into the underlying television program. Example C shows how an advertising message my change based on the selections of the viewer.

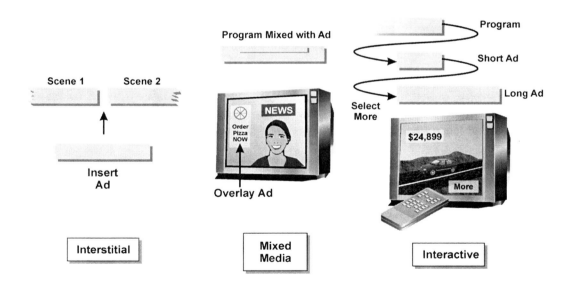

Figure 2.2, Types of IPTV Advertising Messages

35

An IPTV billing system option may include presentation control options for advertising messages. Presentation control is the ability to request, setup and control the display and operation of media on a display such as a television along with the placement and operation of screens on the user's display device. Presentation control is necessary to determine where the advertising messages may appear on a display device in relation to other media. Premium spots and media types (image graphics compared to video) may achieve higher advertising billing rates.

Advertising campaigns may need to be monitored and adjusted by the IPTV service provider due to technical and regulatory requirements that determine the timing and position of advertising messages. For example, governments may limit the display of certain types of advertising messages (such as tobacco and alcohol) and some brand logo owners require a distance between other product photo images to avoid association of a specific product with a specific brand.

Purchasing ad space and time can be for a guaranteed rate (paid placement), estimated rate (ratings based) or via bidding options. The cost of advertising can be valued based on the amount of ad space (display size) that is dedicated for an ad on the television or video display.

Paid placement is a marketing program where companies or people pay a fee for a specific location on a media page (such as a television screen or on a web page). The use of a paid placement program for IPTV advertising assures the time and insertion regardless of its actual popularity ranking.

Rating based advertising sales is the offering of services (such as advertising) where the value or cost of the services is determined by the popularity or subscription to the services. Rating based advertising is used for broadcast advertising where the number of recipients and the type of recipients can vary and is determined by a statistical sampling (such as Neilson ratings) after the program has been broadcasted.

IPTV systems are likely to use ad bidding for the insertion of advertising messages. Ad bidding requires a bid management process that can monitor and adjust the bid amounts for the requested insertion and placement of ad messages. An example of ad bidding is the paying for ad listings on search engines such as Google or Yahoo.

Bidding for IPTV systems is likely to occur for particular segment types. For example, it may be possible to separately bid for different age groups. Because IPTV systems can interact with the viewer, it may be possible to determine which segment the viewer belongs to. For example, if each viewer in a household is provided with a login choice so they can customize their interface (TV screen saver) and access their preferred channel list (MTV compared to Discovery Channel), it is likely that the actual viewer and their characteristics can be determined.

Figure 2.3 shows how IPTV ad bidding may work for selling cars. This diagram shows that bidding for advertising messages may occur for particular age groups, income ranges, program types and geographic regions. This example shows that IPTV advertisers may bid for ads that may appear on a variety of programs throughout various geographic regions. The advertiser sets the maximum bid they are willing to offer and a maximum number of impressions may be selected to ensure advertising budgets can be maintained. This example shows that the advertiser may also be able to select if the same ad should be sent to the same person more than one time.

Ad Name	Age Group	Income Range	Program Type	Regions	Ad Repeats	Bid per Impression	Max Impressions
Utility Vehicle	25-39	70k+	Sports	Nationwide	Yes	0.10	10,000
Status Auto	40-54	Any	Entertainment	Nationwide	No	0.18	20,000
Luxury Car	55-69	70k+	Travel	Florida	Yes	0.07	10,000

Figure 2.3, Ad Bidding

Advertising Metrics

Measuring the performance and success of an IPTV advertising program can be accomplished through the use of existing and new types of marketing measurements including ad impressions, ad selections, ad expansions and ad compressions.

To help companies determine the success of their advertising campaigns, advertising reports are tables, graphs or images that may be provided to represent specific aspects of advertising campaigns or the information or data that is created from advertising campaign. IPTV advertising reports may include the number of ad impressions per segment, number of click through selections, the number of ad expansions and the number of ad compressions.

An ad impression is the presentation of an advertising message or image to a media viewer. Ad selections are the clicking or indication that a button or attribute on an advertising message has been selected. Ad selections are indicated by the click through rate (usually in percentage form), which is a ratio of how many selections (red button) or clicks (mouse selections) an advertising message or item within the ad message receives from visitors compared to the number of times the advertising message is displayed. An example of click through rate for an ad button that is clicked 5 times out of 100 displays to visitors is 5%.

IPTV offers the opportunity for interactive advertising, which allows a user to select or interact with an advertising message. This interaction may result in a redirecting of the source of an advertising message to play a longer more informative version of the ad (expanded ads). Viewers may also be able to end an advertising message to return to their media program (compressed ads).

Because IPTV services are coordinated by a system, this allows for the capturing and analysis of detail program viewing actions such as pauses, channel changes, volume up/down, skip and more information.

Rules Based Advertising

Rules based advertising is the commands and policies that are used when providing advertising messages to viewers. Rules based advertising can define what information is presented (only display one time), when the information is presented (time of day), what media it can be associated with (non-adult content), along with other conditional factors. Rules based advertising can be used to highly target advertising campaigns, which increases the value to the advertiser and consumer and should lead to higher rates for each ad insertion.

Overlay Advertising

An overlay ad is a promotional message that is overlaid on top of another media item. An example of an overlay ad is the insertion of a company or brand logo onto a part of a video display. The use of overlay advertising instead of interstitial ads ensures that the viewer cannot skip over advertising messages using digital video recorders.

It is also possible to combine overlay and contextual advertising on television programs. Ad insertion systems can be setup to select ads based on the context of the program (such as reviewing the closed captioning text) and places these ads on top (overlay) of video during the program.

Contextual Advertising

Contextual advertising is a process that uses descriptive or relative (contextual) words to select advertisements or media messages that will be presented to a potential customer. The use of contextual advertising allows ads to be targeted to prospective customers who watch or look for content that has specific types of content or media. This usually narrows the number of potential customers who can see specific advertising messages.

Figure 2.4 shows a sample report that may be generated for IPTV interactive advertising. This example shows that an advertiser has selected to advertise to specific age groups.

Age Group	Impressions	Click Through
5 through 14	0	0
15 through 24	0	0
25 through 39	416	4
40 through 54	347	11
55 through 69	287	1
over 70	0	0

Figure 2.4, IPTV Advertising Reporting

Addressable Advertising

Addressable advertising is the communication of a message or media content to a specific device or customer based on their device or user address. The address of the customer may be obtained by searching viewer profiles to determine if the advertising message is appropriate for the recipient. The use of addressable advertising allows for rapid and direct measurement of the effectiveness of advertising campaigns.

Highly targeted and addressable advertising offers the potential to increase advertising revenue per viewer by a factor of over 20 times while the viewer experience becomes more personalized and highly targeted. Targeted advertising may become the single most profitable opportunity for IPTV.

Implementing addressable advertising involves adding significant amount of network resources. Some of these resources include ad storage, network switching, transmission bandwidth and ad management systems. It is expected that IPTV will gradually evolve to addressable advertising by upgrading traditional ad insertion systems to allow targeting to regions or small geographic areas first.

Personalized Advertising

Personalized advertising is the communication of a message or media content to one or more potential customers that have been adapted or modified to match the interests of the recipients. Personalized ads may be customized with user preferences such as product types, cultural icons and dialects.

Ads can be customized according to various criteria such as on postal code areas and/or the demographics of that region. Ads could be 'tagged' with the name and contact details of the local dealer.

User Profiling

User profiling is the process of monitoring, measuring and analyzing usage characteristics of a user of a product or service. IPTV service offers the possibility for recording and using viewer information to better target services to users.

While IPTV offers the potential to better track viewing habits, it also has the potential to violate privacy issues. IPTV systems will likely take advantage of user profiling to target promotions or for data mining. To take advantage of user profiles, IPTV systems will use profile management systems.

Profile management is systems capture, organization and access control of service usage information that is related to the device, person or companies who has obtained access rights to the media or service.

Ad Telescoping

Telescoping advertisements are extended advertising messages (selected or automatically expanded) from a smaller and/or shorter version of an ad to a larger and/or longer version of an ad. Ad telescoping allows the viewer to immediately obtain more information about a product or service by selecting an interactive option on the advertising message.

Figure 2.5 shows how ads can be expanded using ad telescoping. This diagram shows that a viewer is presented with an ad that can be expanded to provide more information. If the viewer selects the more button, the channel source is redirected to a longer (expanded) advertising message.

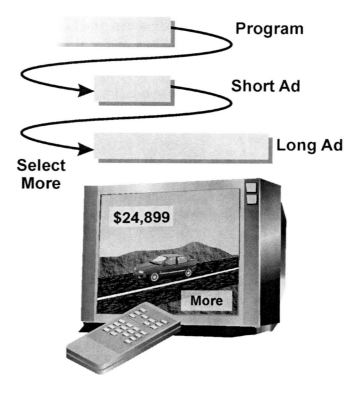

Figure 2.5, IPTV Ad Telescoping

Television Commerce (T-Commerce)

Television commerce (t-commerce) is a shopping medium that uses a television network to present products and process orders. The processes that are used in t-commerce include advanced product offering catalogs (video catalogs), order processing, exchanging of order information between companies in near real-time and the ability to offer multiple forms of payments that may be collected by different companies.

Key issues for IPTV t-commerce billing include transferring accounting records through multiple systems and transferring them between multiple companies that allow for presentation, processing and payment of orders.

An example of IPTV t-commerce is the use of overlay advertising of a consumer item (such as a pizza icon) that appeared during a time period that the viewer had ordered a pizza or a similar product in the past. If the t-commerce system allows the viewer to purchase items and add them to their television (media) bill, this can increase the revenue stream TV service providers. Assuming that many other products can be sold through t-commerce systems, the potential for t-commerce sales can dramatically increase the average revenue per user (ARPU) current generated by television service providers.

To create a t-commerce system, product catalogs are created. Cataloging is the process of identifying media and selecting groups of items to form a catalog. A video catalog is the presenting of items available for selecting or ordering in a video format. Video catalog formats can range from a liner progression of products (such as a television shopping channel) to an interactive video shopping cart that allows users to search and find items.

To integrate catalogs into video programming involves the mixing of media. The mixing of media on the same display (such as the pizza icon) will involve coordination between the television program and the product being offering. This information may include the type of program (e.g. no meat ads during vegetarian documentary programs), suitable time periods for product offers (e.g. no ads during a high action period) and the location and format of the offers (e.g. the product ad not appearing on top of a newscasters face). This means that a lot of descriptive information must be available from both the advertiser and the IPTV service provider.

Figure 2.6 shows how a television program can use mixed media to provide product offers to qualified consumers at specific times in a display location that is noticeable but not intrusive. This picture shows that during a news program, the viewer is presented with a pizza icon from a local pizza restaurant. This example shows that when the user selects the icon, a small window appears with the pizza offer details.

IPTV Business Opportunities

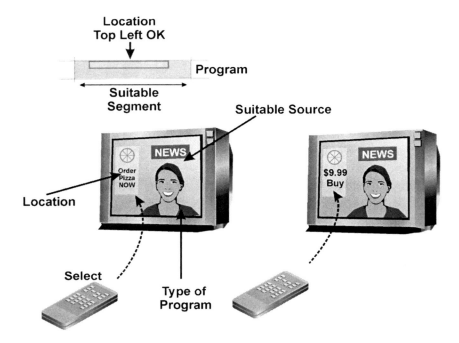

Figure 2.6, Mixed Media Television Product Offer

Vendors may be allowed to directly interact with the products they offer via the t-commerce system. Vendors are likely to know their customers better than IPTV service providers. This means that t-commerce systems will likely include vendor offer management portals. Offer management portals will allow the vendor to add new products, configure their presentation options for the product (e.g. mixed media) and define the product or service offers for specific market segments.

For t-commerce systems, after a viewer has selected a product offer, order processing occurs. Order processing is the defining of terms that are agreeable to the viewer for the acquisition of a product or service. Selected products are placed in a shopping cart for the particular user. Shopping carts are the electronic containers that hold online store items while the user is shipping. The t-commerce shopper is typically allowed to view and change items in their shopping cart until they purchase. Once they have completed the purchase, the items are removed from their shopping cart until they start shopping again.

It is important that the t-commerce system identify the particular user as there can be several users in a household that share an IPTV and each may have orders in progress with a variety of vendors.

Another important part of the t-commerce system is the fulfillment process. Fulfillment is the process of gathering the products and materials to complete an order and shipping the products or initiating the services that were ordered. Depending on the types of products and services offered, IPTV order fulfillment can range from the immediate delivery of media products (such as games or television programs) to the delivery of products over an extended period of time (such as an order of books that has a mix of available and future ship dates).

Customers will likely associate responsibility for fulfilling the order to the IPTV service provider. To reduce the cost of customer care and to avoid potential negative conflicts for unfulfilled products, t-commerce systems may include automated order tracking capability. Order tracking is the ability of a customer, company or other person who is involved with an order to gather information as to the status of the processing of the order.

Figure 2.7 shows a typical scenario of t-commerce order processing. This example shows that a viewer is presented with a product offer (a pizza). This offer is associated with an offer identification code to allow the user to select the offer and to be redirected to an order window. When the user completes the order, the order information is sent to the vendor (the pizza restaurant) where it is confirmed. This diagram shows that order status information may be provided from the vendor to the TV service provider and this information may be used to update the customer about the status of the order (pizza cooking). When the order is complete, the vendor provides information to the TV service provider that the order has been filled to allow the order record to be marked as completed.

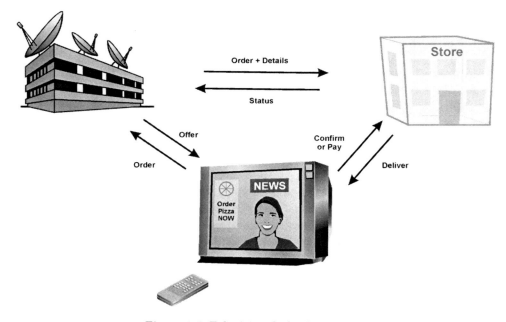

Figure 2.7, Television Order Processing

T-commerce orders can have a variety of payment methods that may need to be recorded in near real time to ensure the vendor and the t-service provider receive payment when they provide products or services.

Payment processing is the tasks and functions that are used to collect payments from the buyer of products and services. Payment systems may involve the use of money instruments, credit memos, coupons, or other form of compensation used to pay for one or more order invoices. T-commerce payment options include payment on the television bill, direct payment collection by the vendor, bill to 3rd party or pay or other payment options.

Figure 2.8 shows payment options for a t-commerce order. This example shows that a t-commerce vendor may receive payment from a t-commerce customer directly by cash or a credit card transaction, the customer may be able to place the order on their television bill or the customer may use a 3rd party such as Paypal to pay for the transaction.

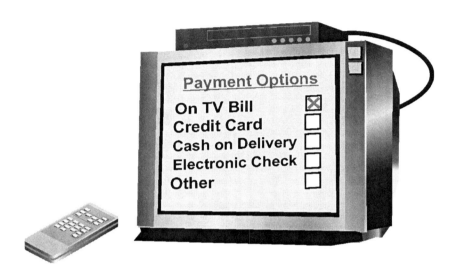

Figure 2.8, T-Commerce Payment Options

To allow multiple companies to process order with multiple TV service providers, a standardized billing communication system is necessary. This billing system will need to transfer a variety of event information including order details, order status and payment information.

IPTV Business Opportunities

Chapter 3

The Key Players in IPTV

Key players are people or companies who participate or have some form of control of an industry, service or product. The key players in the IPTV industry include telephone companies, cable companies, Internet service providers (ISPs), mobile carriers, electric utilities, broadcasters, wireless broadband (WBB) providers and Internet television service providers (Internet TV).

Each type of company has a mix of strengths and areas of weakness when they enter into the IPTV industry. Some of these company types have entered the IPTV industry early (out of necessity in some cases) while others are moving slowly into IPTV (due to lack of awareness).

Telephone Companies

Telephone companies provide and manage access lines between customers and their communication systems. Telephone companies have an extensive amount of switching technology and logistics skill, which is valuable when managing IPTV connections between customer devices and content sources. Telephone companies need to provide high-value IPTV services to customers or they risk financial collapse under a growing load of fixed costs.

Since the late 1990s, customers have been disconnecting from telephone companies. Between 1999 and 2007, some telephone companies have lost more than 40% of their customer base. Unfortunately, the buildings, systems and transmission lines that the telephone companies use and maintain cannot be reduced to match the declining customer base. Telephone companies can offer high-value services (such as IPTV) over existing infrastructure to help recover and increase their service revenues.

Initially, telephone companies have started to deploy IPTV by using DSL transmission over copper lines. Eventually, telephone companies will likely extend their core high capacity communication lines (fiber) directly to the home.

Cable System Operators

Cable system operators obtain and distribute video and multimedia content to customers via controlled communication systems, which allow users to share access to the video content. Cable operators have high-volume business relationships with key content providers that allow them to obtain content at lower cost. Cable system operators will need to convert these systems to IPTV to offer hundreds of thousands of channels so they can maintain competitive advantages with other IPTV operators.

Cable television systems in the early 2000s primarily used simulcast systems that allow for hundreds of shared channels over a mix of fiber and coax distribution lines. Most cable systems have been upgraded to two-way data transmission capability for cable modem and telephone services.

Cable television systems have evolved from companies that retransmit local channels to offering hundreds of linear program channels from a variety of sources. The key cost for most cable companies is content (content cost can be over 50% of revenue), so some cable companies have begun to create their programming content (original programming) to help reduce cost and provide competitive advantages.

Initially, cable system operators may deploy IPTV over the top of cable modem systems. Because the equipment and management of cable modem systems (CMTS) is expensive, will likely evolve their fiber and coax systems to packet data that do not require CMTS equipment.

Chapter 3

Internet Service Providers (ISPs)

Internet service providers (ISPs) supply data communication and information processing services to communication devices. ISPs have experienced programmers and data communication experts, which can rapidly develop, deploy and fix advanced multimedia services. Because ISPs do not have direct high-speed access to customers, they may setup relationships with established or emerging broadband access providers (such as WiMAX systems).

Initially, ISPs may setup systems using technology they are familiar with and have available (low quality video). Over time, ISPs may add multimedia servers that have high quality (e.g. HDTV) content transmission capability. ISPS will need to upgrade their storage capacity and setup content management systems to ingest and serve television content.

Mobile Carriers

Mobile carriers provide wireless communication services to subscribers that can move while they are talking or consuming services. Mobile service providers have experience in providing provide communication services to mobile customers in their own and other systems (roaming) and to bill for usage and advanced services. Because the revenue per customer for voice services are declining and many countries are at market saturation, mobile carriers must begin to provide high value services to maintain and increase their revenue and profits.

Mobile carriers have the capability of sending low and medium speed voice and data signals to customers while they are moving. They track customers through home and visited registrations.

Much of the high-profit services such as music downloading and ring tones have been provided through mobile portals where a relatively high portion of revenues were shared by the mobile carrier. The ability for mobile carriers to block access to non-portal sites decreasing because of competitive pressure (such as the iPhone).

Some of the options available to mobile carriers are to create or use mobile broadcast systems such as DVB-H and MediaFLO to provide a selection of 20 to 50 broadcast television channels. This is a temporary solution as mobile customers are likely to migrate to other broadband wireless systems to get the individual content they desperately desire (such as YouTube and MySpace).

Mobile operators may begin to setup multicast services that allow small groups of mobile users to share popular channels to provide a more individual-like experience. Eventually, mobile carriers will upgrade their systems to multibeam technology, which will allows them to provide individual high quality video channels to many users.

Electric Utilities

Electric utilities provide the necessary energy for customers to use and operate their communication devices. Electric utilities have the capability of sending high-speed data signals directly across power lines within through their power grid directly to devices in the home. IPTV over power systems offers the potential for electric utility companies to rapidly increase their revenues per subscriber at relatively high margins without significant investment.

The ability to provide broadband over power lines (BPL) enables power companies to offer television services. Some power companies are already offering broadband Internet services. It is not a big step to provide IPTV on top of the broadband connections.

The power companies have a key asset "right of way." This gives them the right to install, build or maintain power lines and/or facilities. In addition to linking almost 100% of homes with power wires, electric companies have been installing optical fibers in the lines of the power grid since the 1980s. These fiber optic lines have an amazing amount of transmission capacity, enough to offer IPTV services.

Almost all homes have power lines and all devices require power to operate. The power companies would use their fiber systems to get the television signals close to the home. Once the television signals reach the neighborhood, they travel along the standard power lines to the house at up to 200 Mbps.

Once the power line television signals get to the house, they are re-distributed over the home power distribution system. Once the system is setup, installing a television or viewing device may be as simple as plugging it into an outlet. The television signal will simply follow the power cord into the TV or viewing device.

Broadcasters

Broadcasters obtain and distribute video and multimedia content to customers via wireless communication systems, which allows users to share access to the video content. Broadcasters have setup systems that effectively sell and manage advertising.

Broadcast systems primarily transmit information in one direction from transmitters to viewers. In the 2000s, broadcasters began upgrading their systems from analog to digital, which provides a 5x to 10x improvement on transmission capacity.

IPTV advertising has the potential of increasing ad rates by 10x to 20x while providing even higher returns to advertisers.

Broadcast systems will need to evolve to two-way systems to take advantage of the advanced services available on IPTV. Broadcast IPTV systems are likely to be hybrid systems that use broadcast channels for popular programs and other communication systems (such as Internet connections) for on-demand and Interactive services

Wireless Broadband

Wireless broadband systems can provide high-speed data communications services at data transmission rates of 1 Mbps or higher. Wireless broadband system operators have the capability for rapid deployment and they are aggressively seeking new customers in an industry where many communication competitors are starting to offer broadband services.

Initial deployments of wireless broadband systems such as WiMAX have a limited total transmission capacity of up to 150 Mbps per radio channel. This does not work well for providing broadcast television services to many customers. As a result, wireless broadband system rate plans are typically designed with limited maximum data transmission rates and monthly data transfer limits to deter users from using wireless broadband systems for IPTV services.

Wireless broadband system designs are capable of providing highly focused radio beams providing virtually unlimited capacity. Beam forming systems are already being used in communication systems in India. When wireless broadband systems are upgraded with multibeam capability, wireless broadband systems are likely to become a key player in the IPTV industry.

Internet TV

Internet Television Service Providers (ITVSPs) are companies that provide television or video services that connect through the Internet or other types of broadband data networks. ITVSPs can begin to offer programming globally with relatively low initial equipment costs. The relative simplicity and low initial cost of becoming a global television provider is likely to lure many new content providers to attempt to become Internet Television Service Providers.

Similar to telephone service that is provided through the Internet, ITVSPs setup and provide communication services through the broadband Internet. ITVSPs manage connections between viewing devices such as televisions with adapters (Internet STBs) or multimedia computers and media sources such as live television programs or on demand programs. Reliable, television quality systems have been demonstrated with data rates of 700 kbps. As broadband speeds continue to dramatically increase, Internet TV services will become more popular and managed IPTV system are likely to migrate to unmanaged systems.

Some of the key challenges for ITVSPs include content management, operations, broadband access and regulatory restrictions. Getting and managing content can be difficult for Internet TV providers as content owners may perceive their content may be copied if it is provided through the Internet (this may not actually be true). Operations management may involve installing and configuring equipment and distributing converter boxes. Broadband access connections may prove to be reliable or companies may restrict or intentionally interfere with the delivery of IPTV signals through their network (such as when competition gets fierce). Regulatory restrictions dramatically vary throughout the world and ITVSPs may provide content that is acceptable in one country that is not acceptable in other countries.

The time to market for traditional network operators can be 1 to 2 years or more. Network operators need to obtain licenses, networks need to be built and systems need to be setup and tested. ITVSPs can bypass much of this process. They may even use hosted systems such as NeuLion (www.NeuLion.com) to manage and distribute their content to reduce the time and cost to market.

Figure 3.1 shows the key types of companies that participate in the IPTV industry and their core competencies in the early 2000s. This table shows the types of companies and their competitive strengths. Telecom companies have switching and logistics strength. Cable television systems are masters of content management. ISPs have information technology skills that can create advanced services. Mobile carriers can use their billing and roaming experience for wide area TV distribution. Electric companies can use their existing infrastructure to deliver broadband IPTV signals to devices that need to be plugged in for power. Broadcast companies can leverage their advertising sales systems. Internet TV companies can offer television services globally with relatively low initial costs.

Telecom	Cable TV	ISP	Mobile	Electric	Broadcast	Internet TV
Switching and Logistics	Content Management	Information Technology	Billing and Roaming	Power Necessity	Advertising	Geographic Area and Low Cost

Figure 3.1, IPTV Company Types

Consumers only care about content they desire to view, not the systems. They want to use any device through any network and get it from any content source. The different types of service providers are evolving their systems to provide this universal solution to them. As a result, each of the IPTV service providers will begin to offer similar products and services. This is likely to lead to conflicts between service providers. Some service providers will choose to fight using legal barriers such as exclusive franchise agreements to provide television services.

Figure 3.2 shows how IPTV systems are likely to evolve into a universal system that can connect a variety of devices through different types of systems to many content sources. Customers can access content on mobile devices, standard televisions or multimedia computers. They will connect through a choice of network types. A media gateway will allow them to obtain content from a variety of content provides such as independent, on demand and network program providers.

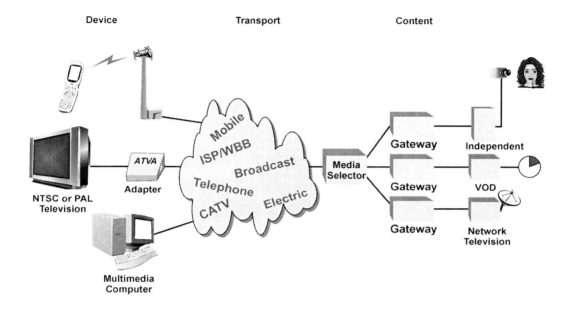

Figure 3.2, Universal IPTV System

IPTV Business Opportunities

Chapter 4

IPTV Content – The Million Channel Need

While it is technically possible for IPTV to provide for easy searching and viewing of millions of channels (similar to the web), millions of channels do not exist yet so where will they come from? While network broadcasters will continue to provide quality programs, new content is being created to fill "The Million Channel Need."

The need for a million television channels comes from marketing and business needs. From a marketing perspective, IPTV service providers are transitioning from a cable television "look alike" to "more than cable" phase. Having thousands of television channels from around the world is a strong differentiating factor. The business reason for offering more channels is to acquire or create popular low cost content that can replace traditional programming sources that can cost 30% to 50% of total revenues.

In 2008, there were over 6 web pages for every human on the planet (more than 39 billion web pages) and yet most consumers had access to only 100 to 300 television channels (or less). In 2008, the typical digital cable television system could provide between 700 to 1000 standard definition television channels (if every channel is converted to digital). Most of the programming for these television systems comes from network feeds (e.g. ABC, NBC and CBS) along with some local and possibly a limited number of international channels.

The initial IPTV systems were cable "look a-likes," which offered traditional programming such as network programming, syndicated programming, local programming and public broadcasting. Some of the new programming sources for IPTV systems include international channels, independent providers, community programming (school, sports and events), corporate programming (public promotional and internal communication) and personal programs (family and friends).

The economic model for IPTV content can be significantly different than traditional broadcasting business models. For IPTV systems, companies pay to develop content and they also pay to host (broadcast) the content as opposed to the broadcast model where TV networks charge for content.

Network Programming

Network programming is the selection of shows and programs that are offered by a television network provider. Network program providers may have affiliate relationships with local content distributors (such as IPTV providers) where the network content is provided at low cost or free if some of the ad spots are held back for the network to sell.

Original Programming

Original programming is content that is owned, developed and controlled by a network operator who provides the media to its viewers. Common forms of original programming include news, documentaries, education and other programming that is created by or for the network operator.

While the creation of original programming may reduce the cost of other programming content (such as network providers), it may still involve the payment of fees or royalties for the use of brands, actors or other images. It may be also be possible to sell original programming through content aggregators turning a cost center into a profit center.

Sponsored Content

Sponsored content is specialty programs or media that are paid for by people or companies who are looking to promote a specific solution or to develop a mailing list of the viewers who watch the content (similar to sponsored Webinars). Sponsored content may be provided on an individual basis where users are required to register or accept the sponsored viewing terms such as sending the contact details of the viewer to the sponsor company so additional details (product brochure) can be sent.

Community Content

Community content programming is media that is created and managed by members of a community or a group that can be viewed by others who are interested in community content. Examples of community content include school, sports and local events that members of a community have an interest in. Community members are commonly interested in assisting in the creation, management and delivery of the community content with or without direct compensation.

Community content programs have been difficult to provide via traditional broadcast television systems because the cost of production can be relatively high in comparison to the number of viewers. It is possible to reduce the cost of content acquisition through the use of volunteers to produce community content. Community members are commonly interested in assisting in the creation, management and delivery of the community content with or without direct compensation.

An example of how effective community content can be was demonstrated by a test project conducted by the company SeaChange. At the Profitability Assurance conference in 2005, SeaChange discussed that local programming including sports events was offered to a community for a free trial period. After the trial period, approximately ¼ of the community members paid to continue access to local programming and the average viewer spent a significant amount of time watching community programming.

Another example of community content programming is the dual function media player developed by Mizu Design. This media player allows the viewer to receive programming and to upload content for community channels. For content that is submitting to a station management system, it is reviewed, edited and scheduled for playout by a community station manager who ensures only acceptable content is available for viewing.

Independent Content

Independent content providers are companies that develop, manage or distribute programming content to virtually any company that wants to broadcast their programming. These companies often specialize in providing programming that is unique and very targeted when compared to network content providers.

An example of an independent content provider is Grid-TV (www.Grid-TV.com). Grid-TV manages 10,000 channels of various categories such as medical television stations. Grid-TV sends more than 130 medical television stations to broadcasters around the world. There are many other independent companies that produce and or distribute content that is available to IPTV service providers.

Shared Content

Shared content is the creation of media by multiple participants. An example of shared content is a blog where group of people participate in message activities that are related to specific subjects.

Wiki TV

Wiki TV is the creation of TV programs by many contributors. WikiTV contributors may include artists, animators, script writers and other people who are motivated to create and to get recognition for their skills.

A Wiki is an application that allows multiple people to easily and quickly contribute (Wiki means Fast in Hawaiian) to a common content base. For Wikis to work, there has to be a pool of qualified contributors, a perceived need or desire for the contributors to participate and a system that enables the management of the Wiki.

There is an abundance of contributors such as artists, animators and event directors that want to get recognition for their skills and ability to perform. It is likely that people will work together to produce animations, audio programs and scripts. It is also likely that they may make many of these available for little or no cost.

There is a growing need for new content from the many outlets for television programming; IPTV, mobile video and broadband TV. The production of quality content in Hollywood is typically very high due to actor's guilds, writer's guilds, unions, directors and structured production processes.

The missing link seems to be wiki web portals and communities to make it happen. When it happens, just like Wikipedia's success, it is likely to rapidly grow to produce cost effective (and possibly well developed) content. For those in Hollywood that can take advantage of Wikis, there is a tremendous opportunity for amazing low cost content production. For those who choose to ignore it, good luck.

Local Programming

Local programming is the selection of shows and programs that are offered by a local television network provider. An example of a local program is a news program that is created and broadcasted by a local broadcaster.

International Programming

International television channels are program sources that originate in other countries. Each developed nation typically has several hundred unique linear television channels. The offering of international programming can be very effective at targeting ex-patriots, which can be a relatively high-percentage of the population in some countries.

As the deployment of IPTV systems accelerate, the demand for international television channel distribution is increasing. Many IPTV systems are offering international programming channels as a way to differentiate IPTV systems from existing broadcast, CATV and satellite systems. While access to these channels may satisfy foreigners living in other countries, it opens up a significant number of potential challenges. These differences can include regulation, access controls, language variations, time offset, media formats, program guides, advertising and content licensing

Regulatory Differences

Countries have different regulations for media broadcasters regarding violence, gambling, product placement and adult content. The regulations can also vary between live, linear and on-demand programming. Because different systems can push or pull media in different ways, a precise definition of the differences between the types of programming is needed. Currently some of the media regulations are defaulting of the "country of origin" regulations. Eventually, content distributors and broadcasters may need to obtain audio/visual licenses from national authorities.

Some countries (such as South Korea) impose product placement restrictions. Product placement restrictions are supposed to remove potential product bias (preferences) or associations that may be a result of the audience seeing a specific brand or product that is used in a movie or television show.

Another example of content restriction is not allowing gambling transactions (as is the case in the United States) to be conducted on television. Muslim countries tend to have restrictions on nudity and violence and Canada is very restrictive on violence in video programming.

Because IPTV is a relatively new medium, it is not well monitored yet. This means that IPTV service providers may be able to distribute the restricted programming without being detected. However, as governments begin to monitor IPTV programming, penalties and significant fines can result. Content aggregators can help to identify and filter restricted media avoiding this potential risk.

Program Access Controls

Program access controls are the processes that viewers must take to obtain access to content. Access controls are typically necessary for some types of content such as adult content or politically sensitive materials. Many industries have already taken self-regulatory initiatives to delay or reduce the impact of government regulations for access controls. Access controls may need to be applied to multiple channels of distribution such as a television broadcast that is delivered through radio broadcasting, cable television and web streaming.

Language Variations

While international television channels may be primarily consumed by foreign nationals who want to obtain the programming in its original language, support for multiple languages may be necessary or required by regulatory agencies.

Language variations are media programs (such as movies or television shows) that contain audio in different languages. Language variations may be created by language dubbing. Dubbing is the insertion of audio information in another language into a media program without affecting the other forms of media (such as a video tape). Language dubbing may be required in certain countries.

Time Offset

Time offset is the difference between the local broadcast time and the time in the country where the international program is being transmitted. The time offset can result in programs from one program time category (such as evening programming) being viewed in another country during time periods that may not be appropriate (such as daytime programming). To overcome the time offset, live programs may be offered in both live and on-demand formats. Time offset may be applied to ensure that the programs are transmitted during approved regulatory time windows (such as programs that may contain nudity only being shown after a certain time).

Media Formats

International programming may be distributed through several types of media communication systems including broadcast TV, mobile TV, satellite, CATV, IPTV and Internet TV. Some broadcasters transmit their programming over multiple platforms. The distribution platform may require changes to the content, which can change the creative component of the media (such as screen aspect ratio).

Program Guides

Program description data should accompany content so that program guides can be created. International programs may not have metadata associated with them, they may use different types of categories or metadata formats and they may be in different languages.

Advertising

While international television channels may be sold on a subscription basis, ad insertion is likely to become the main revenue source for media broadcasters. The advertising messages from the country of origin are likely to be replaced by ads in the country or local area where the programs are viewed. To provide advertising services, broadcasters need to know the ad slots and have the ability to place local ad inserts into the media and they have to be able to sell advertising services.

IPTV companies are using the same marketing and ad management that broadcasters, cable and satellite television companies use, and the complex content licensing models used by broadcasters are not being used by IPTV operators. It is likely that ad management and content licensing systems will get very complex for IPTV systems very soon.

To be able to replace, insert or modify advertising messages, the local broadcaster should have advertising rights.

Company Programming

Company television programming (enterprise TV) is media that is created and managed for or by a company. Company television may be produced for the public and/or for internal communication purposes. Public company television channels may provide information about products, services or applications of the products or services that are of interest to the public. Internal ("in-house") company television programs may be used to provide employees with educational and company specific information (such as the location of a company meeting or party). Company television programs may be distributed to monitors within company buildings or for distribution to multimedia computers that are only accessible by employees, vendors or others who are provided with access.

One of the benefits of company television programming that is made available to the public is self-promotion. The programs are likely to contain products and services offered by the company (product placement).

Personal Programming

Personal television programming is the creation, management or distribution of content for viewing by authorized people (such as friends and family). Personal television channels allow individuals to upload their own content for viewing by people they specifically authorize. Examples of personal media channels include www.myspace.com and www.Veoh.com.

Personal media channels allow viewers to create their own television channel, upload their content such as pictures and videos and share their content with other IPTV viewers. IPTV systems provide access to television channels throughout the world. Some of the more popular global television channels that are available on IPTV include news channels, business channels and music television. IPTVs may provide access to interactive media such as games, chat rooms and e-commerce shopping. Governments and public groups have begun providing real time video access to public sources such as courtrooms, popular public places and public web cams. Private video sources are provided by companies or people who are willing to provide and pay to have their signals available on television. Some common private television channels include religious groups, education sources and sports channels.

An example of how a personal media channel (PMC) may be used for IPTV is the control and distribution of mixed media (such as digital pictures and digital videos) through a personal television channel to friends and family members. An IPTV customer may be assigned a personal television channel. The user can upload media to their personal media channels and allow friends and family access to pictures and videos of family members and gatherings via their IPTVs.

Figure 4.1 shows how personal media channels allow other viewers to be given access to specific types of personal media on their IPTV. This example shows how an IPTV user "Bob" is uploading pictures from a party to his personal media channel 9987. People on his list of viewers who also have IPTV

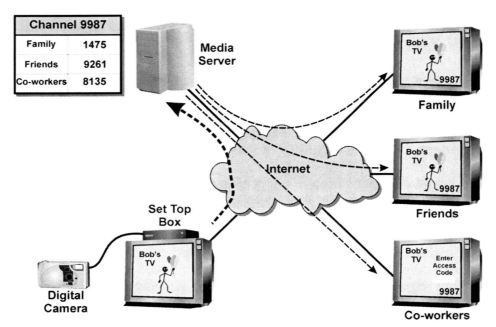

Figure 4.1, Personal Media Channels

service can select Bob's television channel 9987 and see the pictures. This example shows that the personal media channel can have restrictions on who can view the personal media channel.

Content Partners

A content partner is a person, company or organization that owns the rights to content and provides the content along with content rights to distribute the content. Partner management allows administrators to setup accounts for content providers. Content partners can provide content that is branded with their channel or service logos. Brandable content is the ability to create an awareness of a company (corporate brand), product (product brand) or service (service brand) to people or companies.

Content partner accounts contain a variety of terms and provisions for the use, distribution and payment for media programs. Content partners may provide content on a fixed fee or revenue sharing basis. Fixed fees may be charged on a per channel or per subscriber basis. Revenue sharing may divide revenue received from the customers on a subscription and/or on a per view basis.

Wholesale On-Demand

Wholesale on demand is the providing of programs that broadcasters or distributors can dynamically request and receive. Wholesale on demand systems may be used by broadcasters and distributors to add on demand services to networks that are not capable of providing on demand services. Wholesale on demand systems enable almost any broadband network operator to distribute on demand programming without significant investments.

Using wholesale on demand networks allows the cost of on demand network equipment and operations to be shared across multiple operators. In addition to the cost of equipment that is needed to provide on demand services, the cost of acquiring and managing the rights to content acquisition (e.g., movies) can be reduced.

IPTV Business Opportunities

Using wholesale on demand systems, network operators only need to add software in the set-top box (the client), which can communicate with the wholesale on demand system. This software can be unique to the system (proprietary) or it can use open systems.

Figure 4.2 shows some of the existing and new types of IPTV content providers. This diagram shows that IPTV content includes traditional television content sources such as movies, news services, sports, education, religious and other forms of one-way information content. New types of IPTV content include personal media, global television channels, interactive media, public video sources and private video sources.

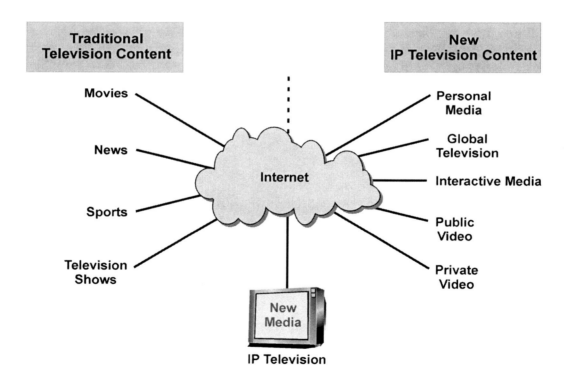

Figure 4.2, IPTV Media Content Providers

Content Management

A content management system identifies, categorizes and manages the storage and distribution of content. Identification of content for an IPTV system can be complicated as there are a variety of media identification codes that are used for different types of media such as media files, streaming media or images.

A content provider may also provide access to live or continuous video sources (e.g. a linear television channel). To provide real time streaming, the video device is instructed to connect to an IP video gateway. The IP video gateway converts or adapts the video source into an IP video stream that the device can receive and display.

Another aspect of content management is the need for the standardization of content classifications. Broadcast media typically includes metadata that describes the content. These descriptions are commonly changed by local broadcast service providers to meet the cultural preferences of their viewers. For example, a program that is classified as a "Thriller" in one geographic area may be classified as a "Suspense" movie in another area. The changing or use of non-standard content management classifications can result in the wrong billing rates or the potential delivery of content to unauthorized viewers (such as the delivery of adult content to minors).

Another potential challenge for content management in IPTV systems is that the characteristics, rights and value of content can change over time. For example, a breaking news story may be given priority display and higher than average advertising rates may be charged when the media is viewed.

Content Lifecycle

Content lifecycle is the progression of content from its concept stage to its end of use or destruction. A typical content lifecycle for media includes the concept, development, production, packaging, distribution and repurposing.

The life of content that is distributed in IPTV systems can vary. Content that may be desirable for long time period (such as a popular television sitcom) is said to have a long tail. Content that is desirable for a brief period is called short tail content.

Long Tail Content

Long tail content is programs or media that is viewed or desired by a small group of people over a relatively long period of time. Long tail typically refers to the statistical distribution of valuable products or content where a majority of usage or distribution occurs at the beginning of a process or product offering. More recently, long tail content has become associated with the re-distribution of media into new channels such as to IPTV systems. With the cost effective distribution offered by IPTV systems, old content (such as old movies) can be given new life (increased value).

Short Tail Content

Short tail content is programs or media that is primarily viewed or desired by a large group of people when it initially is released with a much lower interest and viewing level that occurs over a relatively long period of time. An example of short tail content is a weather report.

Flat Tail Content

Flat tail content is program media that has a relatively consistent viewing demand over a period of time. Flat tail content is programs such as educational or personal development programs that provide enriching information to viewers when they need that information. As time passes, additional people develop need of this information, which results in the continuous demand for flat tail content. An example of flat tail content is an instructional painting lesson (e.g. the "Painting in Watercolors") or a home repair show (e.g. "How to Install Tile").

The key challenge of providing long tail and flat tail content on traditional television distribution systems (such as broadcast television and video stores) is the limited number of programs that can be sent through these channels. Because IPTV systems can offer virtually an unlimited number of television channels, IPTV systems are an ideal choice to distribute a large number of these programs, which is an excellent way for IPTV service providers to uniquely define their services.

Metadata

Metadata is information (data) that identifies and describes the attributes of media or data. Metadata or meta-tags are commonly used in broadcast systems to enable program managers to find, select and setup programming schedules.

Metadata can contain the basic descriptive information (core metadata) such as title, duration and media encoding formats. Metadata may also contain additional information such as viewing rights, costs and scheduling limitations (applied metadata). Metadata may also be used to describe the business requirements for the media such as cost, usage time, authorized types of users and other usage rights restrictions (transactional metadata).

A key opportunity for IPTV business is metadata management. Metadata management is needed to identify, describe and apply rules to the descriptive portions (metadata) of content assets. Because IPTV systems are likely to have thousands or even millions of channels and programs, standardizing and automating metadata is very important.

Content Workflow

Content workflow is the process of assigning, creating and managing the creation or distribution of content. IPTV systems need workflow systems that can manage content projects, schedule production and editing tasks and provide management tools.

IPTV Business Opportunities

Chapter 5

How IPTV and Internet Television Systems Work

Understanding the basics of how IPTV and Internet television service works will help you make better choices and may help you to solve problems that can be caused by selecting the wrong types of technologies, equipment and services.

Digitization - Converting Video Signals and Audio Signals to Digital Signals

A key first step in providing IPTV service is converting the analog video signals into a digital form (if they are not already in digital form) and compressing this digitized information into a more efficient form.

Digitization is the conversion of analog signals (continually varying signals) into digital form (signals that have only two levels). To convert analog signals to digital form, the analog signal is sampled and digitized by using an analog-to-digital (pronounced A to D) converter. The A/D converter periodically senses (samples) the level of the analog signal and creates a binary number or series of digital pulses that represent the level of the signal.

Analog signals are converted into digital signals because they are more resistant to noise (distortion) and they are easier to manipulate than analog signals. For the older analog systems (continuously varying signals), it is not easy (and sometimes not possible) to separate the noise from the analog signals. Because digital signals can only have two levels, the signal can be regenerated and during this regeneration process, the noise is removed. Television signal digitization involves digitization of both the audio and video signals.

Figure 5.1 shows the basic process used to digitize images for pictures from analog video. The image is scanned line by line from the top to bottom. For color video, the image is scanned into lines where each contains intensity (brightness) and color information. This example shows that each line is periodically sampled and converted into digital equivalent levels. This example shows that analog signals can have 256 levels (0-255) and that this can be represented by 8 bits of information (a byte). One byte of information represents the intensity and one byte of information represents the color.

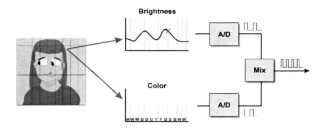

Figure 5.1, Video Digitization

Digital Media Compression – Gaining Efficiency

Uncompressed digital video signals are approximately 270 Mbps for standard definition (SD) video and 1.5 Gbps for high definition (HD) formats. These signals require digital media compression (data compression) so they can be efficiently transmitted to users at 2 Mbps to 10 Mbps.

Chapter 5

Digital media compression is a process of analyzing a digital signal (digitized video and/or audio) and using the analysis information to convert the high-speed digital signals that represent the actual signal shape into lower-speed digital signals that represent the actual content (such as a moving image or human voice). This process allows IPTV service to have lower data transmission rates than standard digital video signals while providing for good quality video and audio. Digital media compression for IPTV includes digital audio compression and digital video compression. The typical ratio of digital compression for IPTV systems ranges from 50:1 (MPEG-2) to 100:1 (MPEG-4).

Figure 5.2 demonstrates the operation of the basic digital video compression system. Each video frame is digitized and then sent for digital compression. The digital compression process creates sequence frames (images) that start with a key frame. The key frame is digitized and used as reference points for the compression process. Between the key frames, only the differences in images are transmitted. This dramatically reduces the data transmission

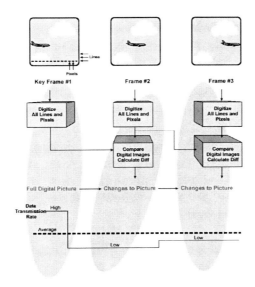

Figure 5.2, Digital Video Compression

rate to represent a digital video signal as an uncompressed digital video signal requires over 270 Mbps compared to less than 4 Mbps for a typical digital video disk (DVD) digital video signal.

Sending Packets

Sending packets through data systems involve routing them through the network and managing the loss of packets when they can't reach their destination.

Packet Routing Methods

Packet routing involves the transmission of packets through intelligent switches (called routers) that analyze the destination address of the packet and determine a path that will help the packet travel towards its destination.

Routers learn from each other about the best routes for them to select when forwarding packets towards their destination (usually paths to other routers). Routers regularly broadcast their connection information to nearby routers and they listen for connection information from connected routers. From this information, routers build information tables (called routing tables) that help them to determine the best path for them to forward each packet to.

Routers may forward packets towards their destination simply based on their destination address or they may look at some descriptive information about the packet. This descriptive information may include special handling instructions (called a label or tag) or priority status (such as high priority for real time voice or video signals).

Figure 5.3 shows how blocks of data are divided into small packet sizes that can be sent through the Internet. After the data is divided into packets (envelopes shown in this example), a destination address along with some description about the contents is added to each packet (called in the packet header). As the packet enters into the Internet (routing boxes shown in this

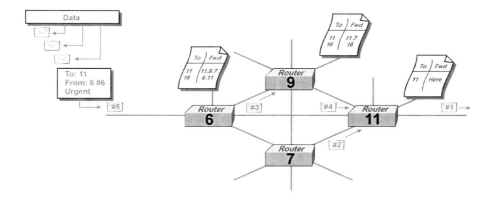

Figure 5.3, Packet Transmission

diagram), each router reviews the destination address in its routing table and determines which paths it can send the packet to so it will move further towards its destination. If a current path is busy or unavailable (such as shown for packet #3), the router can forward the packets to other routers that can forward the packet towards its destination. This example shows that because some packets will travel through different paths, packets may arrive out of sequence at their destination. When the packets arrive at their destination, they can be reassembled into proper order using the packet sequence number.

Packet Losses and Effects on Television Quality

Packet losses are the incomplete reception or intentional discarding of packets of data as they are sent through a network. Packets may be lost due to broken line connections, distortion from electrical noise (e.g. from a lightning spike) or through intentional discarding due to congested switch conditions. Packet losses are usually measured by counting the number of data packets that have been lost in transmission compared to the total number of packets that have been transmitted.

Figure 5.4 shows how some packets may be lost during transmission through a communications system. This example shows that several packets enter into the Internet. The packets are forwarded toward their destination as usual. Unfortunately, a lighting strike corrupts (distorts) packet 8 and it cannot be forwarded. Packet 6 is lost (discarded) when a router has exceeded its capacity to forward packets because too many were arriving at the same time. This diagram shows that the packets are serialized to allow them to be placed in correct order at the receiving end. When the receiving end determines a packet is missing in the sequence, it can request that another packet be retransmitted. If the time delivery of packets is critical (such as for packetized voice), it is common that packet retransmission requests are not performed and the lost packets simply result in distortion of the received information (such as poor audio quality).

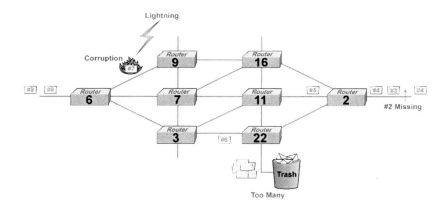

Figure 5.4, Packet Losses

Packet Buffering

Packet buffering is the process of temporarily storing (buffering) packets during the transmission of information to create a reserve of packets that can be used during packet transmission delays or retransmission requests. While a packet buffer is commonly located in the receiving device, a packet buffer may also be used in the sending device to allow the rapid selection and retransmission of packets when they are requested by the receiving device. Packet buffering is commonly used in IPTV systems to overcome the transmission delays and packet losses that occur when viewing IPTV signals.

A packet buffer receives and adds small amounts of delay to packets so that all the packets appear to have been received without varying delays. The amount of packet buffering for IPTV systems can vary from tenths of a second to tens of seconds.

Figure 5.5 shows how packet buffering can be used to reduce the effects of packet delays and packet loss for streaming media systems. This diagram shows that during the transmission of packets from the media server to the viewer, some of the packet transmission time varies (jitter) and some of the packets are lost during transmission. The packet buffer temporarily stores data before providing it to the media player. This provides the time necessary to time synchronize the packets and to request and replace packets that have been lost during transmission.

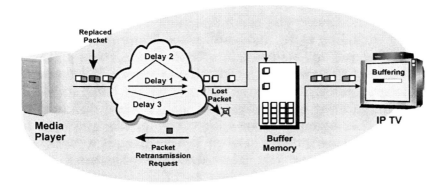

Figure 5.5, Packet Buffering

Converting Packets to Television Service

IPTV data packets are converted back into television signals via media gateways. Media gateways may interconnect IPTV service to a television network (such as a hotel television system) or they may convert the signals directly to a television signal format (such as a NTSC or PAL analog television signals).

Gateways Connect the Internet to Standard Televisions

A television gateway is a communications device or assembly that transforms audio and video that is received from a television media server (IPTV signal source) into a format that can be used by a viewer or different network. A television gateway usually has more intelligence (processing function) than a data network bridge as it can select the video and voice compression coders and adjust the protocols and timing between two dissimilar computer systems or IPTV networks.

Figure 5.6 shows how a media gateway connects a television channel to a data network (such as the Internet). This example shows that the gateway must convert audio, video and control signals into a format that can be sent through the Internet. While there is one communication channel from the gateway to the end viewer, the communication channel carries multiple media channels including video, audio and control information. The gateway first converts video and audio signals into digital form. These digital signals are then analyzed and compressed by a coding processor. Because end users may have viewers that have different types of coders (such as MPEG and AAC), the media gateway usually has available several different types of coding devices. This example shows that the media gateway receives requests to view information (the user or network sends a message to the media gateway). The gateway may have a database (or access to a database) that helps it determine authorized users and the addresses to send IPTV signals.

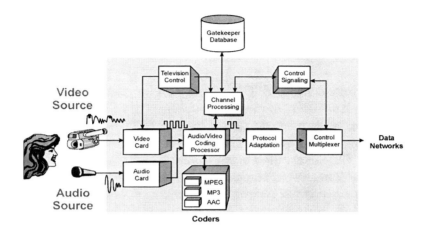

Figure 5.6, IPTV Gateways

Managing the Television Connections

Gatekeepers control the setup, connection, feature operation and disconnection of television channels connected through the data network. Gatekeepers can be owned and operated by private companies or public service providers such as an Internet Television service provider (ITVSP). IPTV systems manage the downloading or streaming of IPTV signals to the consumer and may manage the selection (switching) of the media source.

Switching (Connecting) Media Channels

Gatekeepers are computers that maintain lists of the IP addresses of customers and gateways, process requests for calls and features and coordinate with the gateways that convert IPTV signals to standard television formats. Gatekeepers perform access control, address translation, services coordination, control signaling coordination and bill record recording.

IPTV Business Opportunities

IPTV systems may sets up connections directly between IPTVs or IP set top boxes and media servers or it may use a video switching system to connect the viewer to one of several available media sources. When the media connection is setup directly between the media server and the viewer, this is known as soft switching.

Figure 5.7 shows how a basic IPTV system can be used to allow a viewer to have access to many different media sources. This diagram shows how a standard television is connected to a set top box (STB) that converts IP video into standard television signals. The STB is the gateway to an IP video switching system. This example shows that the switched video service (SVS) system allows the user to connect to various types of television media sources including broadcast network channels, subscription services and movies on demand. When the user desires to access these media sources, the control commands (usually entered by the user by a television remote control) are sent to the SVS and the SVS determines which media source the user desires to connect to. This diagram shows that the user only needs one video channel to the SVS to have access to virtually an unlimited number of video sources.

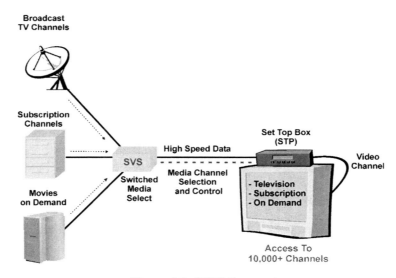

Figure 5.7, IPTV Connection

Multiple IPTVs per Home

Each household may have several users that desire to watch different programs. This requires the bandwidth to be shared with each individual IPTV.

Households may have a combination of several multimedia computers, set top boxes or IPTVs in each home. When viewers are watching television channels (different channels), the bandwidth of each IPTV signal must be added.

Figure 5.8 shows how much data transfer rate it can take to provide for multiple IPTV users in a single building. This diagram shows 3 IPTVs that require 1.8 Mbps to 3.8 Mbps to receive an IPTV channel. This means the broadband modem must be capable of providing 5.4 Mbps to 11.4 Mbps to allow up to 3 IPTVs to operate in the same home or building.

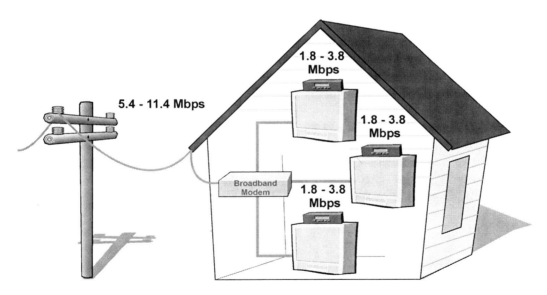

Figure 5.8, IPTV Multiple Users

Transmission

IPTV channel transmission is the process of transferring the television media from a media server or television gateway to an end customer. IPTV channel transmission may be exclusively sent directly to specific viewer (unicast) or it may be copied and sent to multiple viewers at the same time (multicast)

Unicast

Unicast transmission is the delivery of data to only one client within a network. Unicast transmission is typically used to describe a streaming connection from a server to a single client.

Unicast service is relatively simple to implement. Each user is given the same address to connect to when they desire to access that media (such as an IPTV channel). The use of unicast transmission is not efficient when many users are receiving the same information at the same time because a separate connection for each user must be maintained. If the same media source is access by hundreds or thousands of users, the bandwidth to that media server will need to be hundreds or thousands of times larger than the bandwidth required for each user.

Figure 5.9 shows how IPTV systems can delivery the same program to several users using unicast (one-to-one) channels. This example shows that each viewer is connected directly to the media server. Because each viewer is receiving 3 Mbps, the media server must have a connection that can provider 9 Mbps (3 Mbps x 3 viewers).

Chapter 5

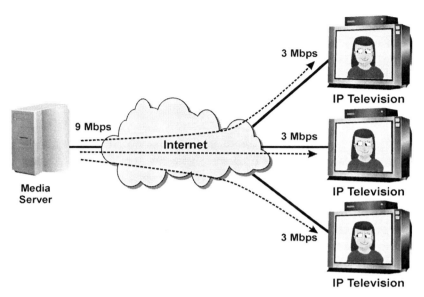

Figure 5.9, IPTV Unicast Transmission

Multicast

Multicast transmission is a one-to-many media delivery process that sends a single message or information transmission that contains an address (code) that is designated that allows multiple distribution nodes in a network (e.g. routers) to receive and retransmit the same signal to multiple receivers. As a multicast signals travels through a communication network, it is copied at nodes within the network for distribution to other nodes within the network. Multicast systems form distribution trees of information. Nodes (e.g. routers) that copy the information form the branches of the tree.

The use of multicast transmission can be much more efficient when the same information is sent to many users at the same time. The implementation of multicast systems is generally more complex than unicast systems as more control is required to add and remove members of multicast groups. Multicast recipients generally submit requests to a nearby node within a multicast network to join as part of an active multicast session.

IPTV Business Opportunities

For multicast systems to operate, nodes (routers) within the network must be capable of multicast sessions. Because of the complexity and cost issues, many Internet routers do not implement multicast transmission. If the multicast network is controlled by a single company (such as a DSL or cable modem data service provider), all the nodes within the network can be setup and controlled for multicast transmission.

Figure 5.10 shows how an IPTV system can distribute information through a switched telephone network. This example shows that end users who are watching a movie that is initially supplied by media center that is located some distance and several switches away from end users (movie watchers). When the first movie watcher requests the movie, it is requested from the telephone end office. The telephone end office determines that the movie is not available in its video storage system and the end office switch requests the movie from the interconnection switch. The interconnection switch also

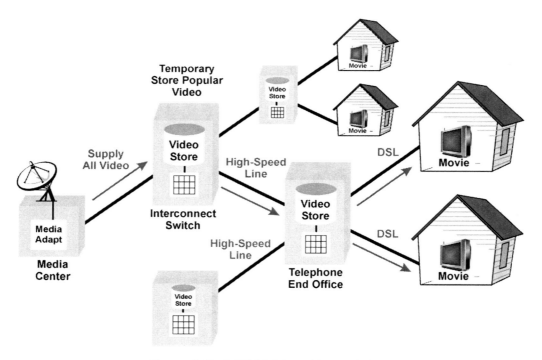

Figure 5.10, IPTV Multicast Transmission

determines the movie is not available in its video storage system and the movie is requested from the distant media source. When the movie is transferred from the media center to the end customer, the interconnecting switches may make a copy for future distribution to other users. This program distribution process reduces the interconnection requirements between the switching distribution systems.

Channel Selection

IPTV channel selection is the process of finding and connecting to an IPTV data address (IP address) so it can receive and decode a television or media channel.

It may be possible for viewers to direct connect to IPTV channels if they know the URL or IP address (web link). Although the viewer may have the address or the URL of the IPTV channel (a media server), the viewer may not be authorized to connect to the channel at that address. Owners of IPTV media can restrict access to paying customers. Authorization codes are typically pre-established by viewers or companies that provide IPTV services to viewers.

IPTV service providers provide a selection screen or device that allows users to find and select IPTV channels. Because it is difficult for viewers to remember or organize URLs and IP addresses, channel selection screens usually have more descriptive information such as channel numbers, network names and show titles. While it is possible to have IPTV service that uses channel numbers that are identical to standard cable television systems, IPTV service providers (ITVSPs) offer many new ways to find and select television channels.

Program Guide

A program guide is a listing or an interface (portal) that allows a customer to preview and select from possible lists of available content media. Program guides can vary from simple printed directory to interactive filters that dynamically allow the user to filter through program guides by theme, time period, or other criteria.

IPTV service providers usually provide an electronic programming guide (EPG) which is an interface (portal) that allows a customer to preview and select from possible lists of available content media. EPGs can vary from simple program selection to interactive filters that dynamically allow the user to filter through program guides by theme, time period or other criteria. Viewers are also able to connect to IPTV channels through the use of a web link on web pages or through a link that is sent (embedded) in emails.

Figure 5.11 shows some of the different ways a user can find and select IPTV channels. While it is possible for IPTV systems to use channel numbers for the selection of IPTV channels, this example shows that there are several new more effective ways to search and select channels. The user can search for channels by favorites, country, actor name, show title, network provider and category. The user can also select from the channel numbers offered by their IPTV provider.

Figure 5.11, IPTV Channel Selection

Recommendation Engine

A recommendation engine is an application that searches through data files or related listings of information (such as television programs) to find matches to categories or items that the person who is searching or viewing programs is likely to be interested in obtaining or viewing.

A recommendation engine uses a relational metadata server to provide the user with choices that better match the interests and desires of the user. In research gathered by Hillcrest Labs, consumers were 3x as likely to purchase additional content when the navigation system was simplified and enhanced.

Addressable Advertising

Addressable advertising is the communication of a message or media content to a specific device or customer based on their address. The address of the customer may be obtained by searching viewer profiles to determine if the advertising message is appropriate for the recipient. The use of addressable advertising allows for rapid and direct measurement of the effectiveness of advertising campaigns.

A key aspect of addressable advertising is the validation of the viewer. IPTV systems may ask (prompt) the viewer to select their name from a list of registered users in the home when the IPTV is turned on. Because of the advanced features offered by IPTV such as incoming calls/emails and programming guides that remember favorite channels, viewers will typically want to select their programming name. Because the programming name has a profile (preferences), advertising messages can be selected that best match the profile.

The potential revenue for addressable advertising messages that are sent to viewers with specific profiles can be 10 to 100 times higher than the revenue for broadcasting an ad to a general audience. The ability to send ads to a specific number of viewers allows advertisers to set specific budgets for addressable advertising. It also allows the advertiser to test a number of different ads in the same geographic area at the same time.

Figure 5.12 shows how addressable advertising can be used to better match advertising messages to the wants and needs of viewers. This diagram shows that a media program (such as a television show or movie) is being sent to 3 homes where the televisions in each home each have a unique address. When the time for a 30 second commercial occurs, a separate advertising message is sent to each one of the viewers based on the address of the television. This allows each viewer to receive advertising messages that are better targeted to their needs and desires.

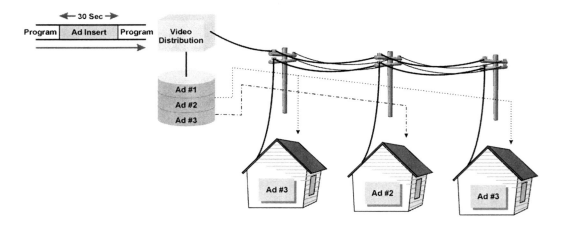

Figure 5.12, Addressable Advertising

Video on Demand (VOD)

Video on demand is a service that provides end users to interactively request and receive video services. These video services are from previously stored media (entertainment movies or education videos).

Chapter 5

VOD can be enhanced by advanced electronic programming guides that maintain a history of previously viewed television or media shows. This allows the viewer to scan through a list of programs that they have not previously viewed.

Figure 5.13 shows how IPTV can allow a viewer to request control the presentation of television programs on demand. This diagram shows that a television on demand viewer can browse through available television channels. In this example, this IPTV service provider informs the viewer of which programs they have already viewed and the length of time each program will run. When the user selects a potential program to view, a short description of that program is shown at the bottom of the screen along with the cost for viewing that particular program.

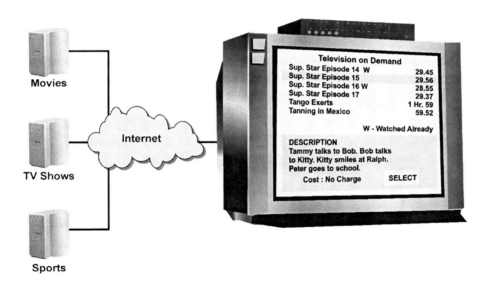

Figure 5.13, Television on Demand

Download and Play

Download and play is a process of downloading a media program (an audio or video file) and then playing it after the file has completely downloaded. A download and play system requires the user to download programs through the Internet before they can begin viewing the programs. This ensures that the programs can be viewed at their encoded quality format (SD, HD) without any quality of service (QoS) issues.

The amount of time that is required to download programs varies based on the speed of the connection (both on the receiving and sending ends) and the media format (standard or high definition video).

Using a broadband connection of 4 Mbps, it takes approximately 15 minutes to download a 1 hour television program in standard definition format. Programs in high definition format take 2 to 4 times longer to download. The typical rule of thumb is 1 hour program divided by data connection speed in Mbps is the download time (in hours). For example, a 1 hour program that is transferred by a data transfer rate of 60 Mbps connection would take 1 minute (1/60th of an hour) to download.

Figure 5.14 shows how a download and play system can allow a user to watch programs that are transferred through a data network (such as the Internet). This example shows that the user selects the media and they

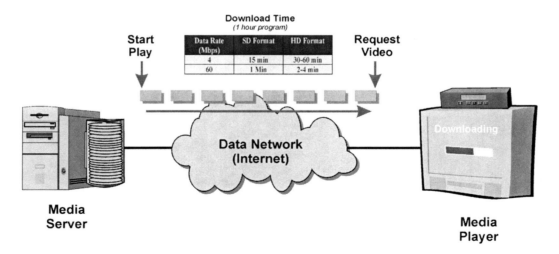

Figure 5.14, Download and Play IPTV

must wait for the program to download before playing. This example shows a set top box that has an 80 GB hard disk that can hold approximately 80 hours of standard definition programs. The table shows that the approximate amount of time for program download if they have a 4 Mbps connection is 15 minutes for programs in standard definition (SD) format and 30 to 60 minutes for programs that are in high definition (HD) format. If the data connection speed is 60 Mbps, the download time per hour is 1 minute for SD and 2 to 4 minutes per hour for HD.

The storage system used in download and play capable set top box can typically hold at least 80 to 100 hours of standard definition (SD) quality television programs. Although it is possible to fill up the download and play STB, content may expire after a period of time (1 to 30 days) and it can be automatically removed from the storage system.

Download systems can be complete self-service systems. Download systems can process and fill orders from viewers without any communication required from the content distributor or access providers. Download equipment may be pre-configured so it is ready to plug into the Internet. When the unit is initially turned on, an account setup screen is required registration and entry of a form of payment.

Streaming

Media streaming is a media transmission method that provides a continuous stream of information that is commonly used for the delivery of audio and video content with minimal delay (e.g. real-time). Streaming signals are usually compressed and error protected to allow the receiver to buffer, decompress and time sequence information before it is displayed in its original format. Streaming media programs are usually not stored locally in the end user device. This reduces the cost of end user devices and it ads a level of security for content owners as the media is not available for copying in the end user device.

IPTV Business Opportunities

Brief delays in streaming transmission can cause the picture to momentarily stop (freeze) or experience significant amount of distortion. To overcome this, progressive downloading can be used. Progressive downloading is the transferring of a file or data in a sequential process that allows for the using of portions of the data before the transfer is complete.

Figure 5.15 shows how to stream movies through the Internet. This diagram shows that streaming allows the media player to start displaying the video before the entire contents of the file have been transferred. This diagram also shows that the streaming process usually has some form of feedback that allows the viewer to control the streaming session and to provide feedback to the media server about the quality of the connection. This allows the media server to take action (such as increase or decrease compression and data transmission rate) if the connection is degraded or improved.

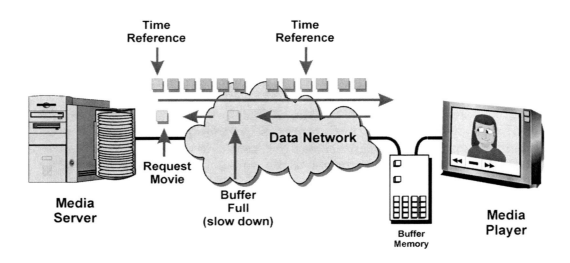

Figure 5.15, Streaming Movies through the Internet

Progressive Downloading

Progressive downloading is the transferring of a file or data in a sequential process that allows for the using of portions of the data before the transfer is complete. Progressive downloading can be used to allow viewers to have instant access to programs. Progressive downloading works if the program that is viewed can be transferred to their device faster than they are viewing it.

Push Video on Demand (PVOD)

Push video on demand (PVOD) is the downloading of video content that is not previously requested by the consumer so that it can be immediately available if the user selects to watch the program. Push VOD usually involves transparent content distribution where the transfer of media or information into a device or system (such as a television set top box) is not visible to the recipient or user.

PVOD systems operators select which programs are likely to be consumed during specific time periods, determine when the network capacity is available for delivery of these programs, identify available storage capacity in the viewer devices (the set top box memory) and coordinate the distribution of the programs. After the programs have been downloaded, the user has instant access to the content. The content is usually stored in encrypted form and the PVOD system coordinates access controls for the usage (viewing) of the stored programs.

Home Network Management

Home network management is the processes that are used to setup and manage premises distribution networks and ensure QoS levels regardless of the devices the consumer attaches to their home networks. IPTV operators may be able to connect to broadband modem gateways that can remotely manage home media networks. IPTV service operators can use these gateways to see and adjust the configuration of the home networks increasing customer satisfaction and reducing the need for truck rolls.

Service Provisioning

IPTV network systems usually provide the viewer with more direct control over television services. IPTV service is typically activated and changed directly through a screen display or Internet web page. Instead of using a customer service representative (CSR) from the television company, the user may be able to setup IPTV services directly. These changes such as service activation and feature addition/deletions can have immediate results.

Activation

Activation is the process of obtaining service after applying for service. If you already have access to a data connection, service activation for services that use the data link for connections (such as Internet Television service) can be instant.

Figure 5.16 shows how it is possible for a user or company system administrator to instantly activate a new IPTV service. In this example, the ITVSP has created a web access page that allows the user to self activate them-

Figure 5.16, IPTV Instant Line Activation

selves. After the user has provided the necessary information such as the billing address and method of payment, account identification codes can be provided manually to the user or they may be automatically entered into the IPTV viewer. The user can then select feature preferences such as preferred television channels and viewer profile.

Conditional Access System (CAS)

Conditional access is a system or service access control process that is used in a communication system (such as a broadcast television system) to limit the ability of users to obtain or use media or services. Conditional access systems can use uniquely identifiable devices (sealed with serial numbers) and may use smart cards to store and access secret codes.

Digital Rights Management (DRM)

Digital rights management is a system of access control and copy protection used to control the distribution of digital media. DRM involves the control of physical access to information, identity validation (authentication), service authorization and media protection (encryption). DRM systems are typically incorporated or integrated with other systems such as content management system, billing systems and royalty management. Some of the key parts of DRM systems include key management, product packaging, user rights management (URM), data encryption, product fulfillment and product monitoring.

DRM systems identify and use devices that it expects to be reliable and authorized for use (trusted devices). Trusted devices are hardware components or software applications that are previously known or suspected to only communicate information that will not alter or damage equipment of stored data. Trusted devices are usually allowed privilege levels that could allow data manipulation and or deletion. Hardware systems and custom made integrated electronic circuits (such as set top boxes) are not as easy to modify as software application programs.

Untrusted devices are hardware components or software applications that are unknown or not validated with a provider of data or information. An untrusted device may require authentication based on some type of user interaction before access is granted.

DRM systems manage secret information ("keys") that is used to identify and modify (decode) media. Key management is the creation, storage, delivery, transfer and use of unique information (keys) by recipients or holders of information (data or media) to allow the information to be converted (modified) into a usable form.

DRM systems must also manage the selecting and coding of media that are in different formats. Different types of devices require different media formats and the control protocols and their various versions (such as SIP) that access these media files may vary.

DRM systems assign users with specific rights and abilities to access, store and view media. User rights management (URM) is the process of selecting, assigning and terminating (revoking) the authorization of access and use privileges of a user of a product or service. The types of user rights that may be assigned may include a length of time, number of users and the ability to copy or modify the product or service.

DRM systems may validate (authenticate) the identity of the user to ensure they are the person or device that is authorized to request and receive the media. Authentication is a process during where information is exchanged between a communications device (typically a user device such as an IPTV or mobile phone) and a communications network that allows the carrier or network operator to confirm the true identity of the user (or device). The validation of the authenticity of the user or device allows a service provider to deny service to users that cannot be identified. Thus, authentication inhibits fraudulent use of a communication device that does not contain the proper identification information.

To help protect digital information from being used by unauthorized recipients, DRM systems may encrypt the media. The DRM system controls access through the providing of decryption (decoding) keys.

Data encryption is a process of a protecting video and voice information from being obtained by unauthorized users. Encryption involves the use of a data processing algorithm (formula program) that uses one or more secret keys that both the sender and receiver of the information use to encrypt and decrypt the information. Without the encryption algorithm and key(s), unauthorized listeners cannot decode the message. When the encryption and decryption keys are the same, the encryption process is known as symmetrical encryption. When different encryption and decryption keys are used (such as in a public encryption system), the process is known as asymmetrical encryption.

Data encryption of media data may be performed on a group or individual basis. When data encryption is performed on a group basis, the information can be stored in an encrypted form. If the data is performed on an individual basis, the data is typically encrypted immediately prior to transfer to the viewer. How the media is formatted and transferred to the end user is called product packaging. During the product packaging process, the information may be selected (encrypted with a pre-assigned key) or modified (encrypted with a unique key). The transfer of the media file to the recipient is called product fulfillment.

Encryption may be automatically provided between two points on a network. For example, on a cable modem, there is usually encryption between the cable modem and the cable network's connection to the Internet. This is important, as several users on the cable system will have physical access to the signals of other users on the cable network.

DRM systems may also monitor for the unauthorized use and transfer of digital media. Digital media may be identified by file names, media components or embedded digital watermarks. Product monitoring may be performed by searching the web using web crawlers or by locating data sniffers at well known transfer points on the Internet. When unauthorized transfer is observed, action can be taken to inform the media host that they are violating copyright laws and they must remove the content or risk loosing their Internet access authorization (disabling the site).

Chapter 6

IPTV Systems

IPTV systems are composed of viewing devices, television adapters (STBs), premises distribution networks, broadband access systems, headend control and distribution and content acquisition networks.

Viewing Devices

IPTV channels can be viewed on a multimedia computer, standard television using an adapter or on a dedicated IPTV. Multimedia computers have video processing and audio processing capabilities. Television adapters convert digital television signals into standard television RF connections that allow standard televisions to watch IPTV channels. IPTVs are devices that are specifically designed to watch television channels through IP data networks without the need for adapter boxes or media gateways.

Figure 6.1 shows several types of IPTV viewing devices. This diagram shows that some of the options for viewing devices include multimedia computers, television adapters, IPTVs and mobile telephones. Multimedia computers (desktops and laptops) allows some multimedia computers to watch Internet television programs without the need for adapters provided they have the multimedia browsers that have the appropriate media plug-ins. Television adapters connect standard television RF or video and audio connectors to data jacks or wireless LAN connections. IPTVs can be directly connected to data jacks or wireless LAN connections. Mobile telephones that have multimedia capabilities.

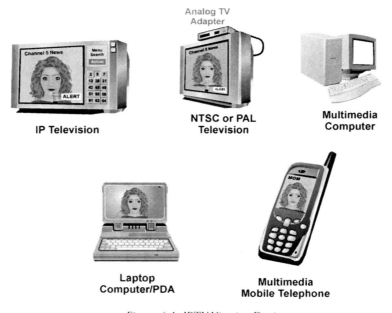

Figure 6.1, IPTV Viewing Devices

Multimedia Computer

A multimedia computer is data processing device that is capable of using and processing multiple forms of media such as audio, data and video. Because many computers are already multimedia and Internet ready, it is often possible use a multimedia computer to watch IPTV through the addition or user of media player software. The media player must be able to find and connect to IPTV media servers, process compressed media signals, maintain a connection and process television control features.

Control of the IPTV on a multimedia computer may be performed by the keyboard, mouse, or external telephone accessory device (such as a remote control) that may be connected to the computer through an adapter (such as an infrared receiver). The media player software controls the sound card, accessories (such as a handset) and manages the call connection.

IPTV signals may be able to be displayed on a multimedia provided it has enough processing power (processing speed) and the necessary media player protocols and signal decompression coders. IPTV signals contain compressed audio and video along with control protocols. These signals must be received, decoded and processed. The processing power of the computer may be a limitation for receiving and displaying IPTV signals. This may become more apparent when IPTV is taken from its small format to full screen video format. Full screen display requires the processor to not only decode the images but also to scale the images to the full screen display size. This may result in pixilation (jittery squares) or error boxes. The burden of processing video signals may be decreased by using a video accelerator card that has MPEG decoding capability.

A media player must also have compatible control protocols. Just because the media player can receive and decode digital video and digital audio signals, the control protocols (e.g. commands for start, stop, and play) may be in a protocol language that the media player cannot understand.

IP Televisions

IPTVs are television display devices that are specifically designed to receive and decode television channels through the Internet without the need for adapter boxes or media gateways. IPTVs contain embedded software that allows them to initiate and receive television through the Internet using multimedia session protocols such as SIP.

An IPTV has a data connection instead of a television tuner. IPTVs also include the necessary software and hardware to convert and control IPTV signals into a format that can be displayed on the IPTV (e.g. picture tube or plasma display).

3 Dimensional Displays (3D Displays)

A three dimensional display is a device that converts (renders) media or data into a format that produces a perception of three dimensions. By filming media with two camera's that are only a few inches apart and then projecting the media in a manner where the viewer will watch each camera's output with only one eye, the illusion by a person is that they can perceive height, width and depth of images. Since depth perception is provided by providing different images to each eye, various techniques are used, such as colored glasses, polarized glasses and other methods that filter the media to provide different image for each eye.

Multiview Lenticular Display

A multiview lenticular display is a device that provides multiple viewpoints using an optical lens that provides several different viewing angles to a single display. When a viewer is located at a viewpoint where they can view different images in each eye, they can perceive image depth. Using multiview lenticular displays allows for the use of multiple viewing points, which is an optical solution. The display screen essentially wears "glasses" that direct two images towards the user. Just like wearing glasses, the "focal point" is set to hit the eye and provide the eye with a sharp image. Similarly, if you move your orientation from the lenses, you lose the image. The lenticular display projects two images that are intended to be viewed by each eye separately in order to create the 3-D image. This implies that you must sit at the same spot front of the 3D display to view media. However, a solution to this challenge is to design a display that has multiple focal points.

To create a display with multiple viewpoints, a display that has a set of lenses on top of the LCD screen is used. Each of the cylindrical lenticules focus pixel images below it to a particular focal point. By installing a display with multiple lenticular lenses, as the viewer moves to different viewing points, they can see different pixels below the lenticules.

Figure 6.2 shows how a multiview lenticular display can provide depth perception and allow multiple viewers to have different perspectives. This example shows that the display provides several viewpoints that display different left and right images. Each of the viewpoints sees different pixels located behind the lenticular lens. As the viewer moves to different points, they can see different display pairs. This allows the viewer to have a different image with depth perception at multiple locations in front of the display without the need for eyewear accessories.

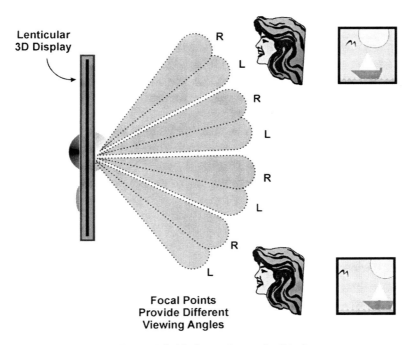

Figure 6.2, Mutliview Lenticular Display

3-D displays are in their infancy but the potential benefits of enabling three-dimensional images on an essentially flat screen potentially can provide some valuable services. Imagine air traffic control screens in 3-D; medical images that provide better information than x-rays at a cost far less than MRIs and of course, gaming, sports viewing and enhanced messaging.

IP Set top Boxes (IP STB)

An IPTV set top box (IP STB) is an electronic device that adapts a broadband data communications signal into a format that can be displayed on a television. The basic functions of an IP set top box include connecting to the broadband data connection, separating, decoding and processing media and interfacing the media and controls with the user.

IP STB adapters are devices that are designed to convert digital broadband signals into analog television formats (e.g. NTSC or PAL). Using IP STB, it is possible to use standard televisions for viewing television channels that are sent over data network such as the Internet.

The IP STB may be attached to a network connection such as a wired Ethernet, DSL, fiber optic or wireless connection. It automatically detects if a data connection is available and requests an IP Address. It then registers with the IPTV system and the user can begin selecting and watching live or stored (VOD) television programs.

Sensory Accessories

Sensory accessories range from 3D video eyeglass viewers to network connected sex accessories. A virtual golfing interface from Electric-Spin allows the user to play virtual golf with other players using their smart Launchpad™ accessory that tracks virtually all aspects of the golf swing. In addition to extending IPTV systems using gaming accessories, there is also a merging of IPTV services into networked gaming devices such as the XBOX 360™.

Premises Distribution

A premises distribution system is the interconnection of devices within the premises of a viewer or customer. Premises distribution systems can use copper (wire or coax), optical cable or wireless (radio or IR) to transfer communication of signals within a customer's facility or home. The common premises distribution system used in IPTV include wired LAN, telephone wiring, coaxial cable, wireless LAN, power line distribution and wired LAN.

Wired LAN

Wired LAN systems use cables to connect routers and communication devices. These cables can be composed of twisted pairs of wires or of optical fibers. Wired LAN data transmission rates vary from 10 Mbps to more than 1 Gbps.

When wired LAN systems use twisted cable, the data transmission rate (cable rating) is based on the number of twists in the cable. The higher the number of twists, the higher the maximum data transmission rate and the higher the category rating of the LAN cable.

Wired LAN systems are typically installed as a star network. The star point (the center of thee network) is usually a router or hub that is located near a broadband modem. LAN wiring is not a commonly installed in many homes and when LAN wiring is installed, LAN connection outlets are unlikely to be located near television viewing points.

Telephone Wiring

Telephone wiring premises distribution systems transfer user information over existing telephone lines in a home or building. Telephone data transmission rates vary from 1 Mbps to over 100 Mbps. Telephone line outlets may be located near television viewing points making it easy to connect IPTV viewing devices.

Telephone lines to and from the telephone company may contain analog voice signals, uplink data signals and downlink data signals. These frequency bands typically range up to 1 MHz. Phone line premises distribution systems may use frequencies above the 1 MHz frequency band to transfer signals to telephone jacks throughout the house. Adapter boxes or integrated communication circuits convert the video and/or data signals to high frequency channels that are distributed to different devices located throughout the house. To ensure the phone line signals do not transfer out of the home to other nearby homes, a blocking filter may be installed.

An industry standard for multimedia distribution over telephone lines is managed by the Home Phoneline Networking Alliance (HomePNA). The HomePNA is a non-profit association that works to help develop and promote unified information about a phoneline technologies, products and services. The HomePNA has created the HPNA specification. The HPNA specification defines the signals and operation for data and entertainment services that can be provided through telephone lines that are installed in homes and businesses. The HPNA specification is designed to co-exist with other communication systems including POTS, ISDN and ADSL. More information about HomePNA can be found at www.HomePNA.org.

Coaxial Cable

Coaxial cable premises distribution systems transfer user information over existing coaxial television lines in a home or building. Coaxial cable data transmission rates vary from 1 Mbps to over 1 Gbps . Many homes have existing cable television networks and the outlets, which are located near video accessory and television viewing points.

Coaxial systems are setup as a tree distribution system. The root of the tree is usually at the entrance point to the home or building. The tree may divide several times as it progresses from the root to each television outlet through the use of signal splitters.

Coax lines to and from the cable television (CATV) company may contain analog and digital television and modem signals. Cable television distribution systems use lower frequencies for uplink data signals and upper frequencies for downlink data and digital television signals. Some of the center frequencies are used for analog television signals. These frequency bands typically range up to 1 GHz. Coax premises distribution systems can use frequencies above the 1 GHz frequency band to transfer signals to cable television jacks throughout the house. Adapter boxes or integrated communication circuits convert the video and/or data signals to high frequency channels that are distributed to different devices located throughout the house. To ensure the coax premises distribution signals do not transfer out of the home to other nearby homes, a blocking filter may be installed.

Wireless LAN

Wireless LAN premises distribution systems transfer user information over a WLAN system in a home or building. Wireless LAN data transmission rates vary from 2 Mbps to over 54 Mbps.

Multimedia signals such as television and music are converted into WLAN (Ethernet) packet data format and distributed through the home or business by wireless signals. Some versions of the 802.11 WLAN specifications include the ability to apply a quality of service (QoS) to the distributed signals giving priority to ensure that time sensitive information (such as video and audio) can get through and non-time sensitive information (such as web browsing).

Power Line Wiring

Power line wiring premises distribution systems transfer user information over existing electric power lines in a home or building. Power line data transmission rates vary from 14 Mbps to over 85 Mbps. IPTV appliances typically plug into a power line outlet making power line premises distribution a no-install type of distribution system.

Power line premises distribution standards are being developed by the Homeplug powerline alliance. The Homeplug specification defines the signals and operation for data and entertainment services that can be provided through electric power lines that are installed in homes and businesses. More information about Homeplug can be found at www.Homeplug.org.

Figure 6.3 shows the common types of premises distribution systems that can be used for IPTV systems. This diagram shows that an IPTV signal arrives at the premises at a broadband modem. The broadband modem is connected to a router that can distribute the media signals to forward data packets to different devices within the home such as IPTVs. This example shows that routers may be able to forward packets through power lines, telephone lines, coaxial lines, data cables or even via wireless signals to adapters that receive the packets and recreate them into a form the IPTV can use.

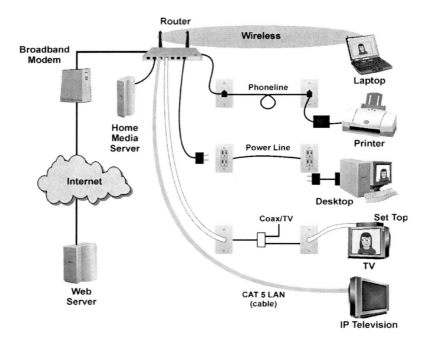

Figure 6.3, IPTV Premises Distribution

Broadband Access Network

Broadband access providers transfer high-speed data to the end users. The type of technology used by broadband access providers can play an important part in the ability and/or quality of IPTV services.

Broadband access systems are networks that can transfer data signals at a rate of 1 Mbps or more. Broadband access systems that can provide data transmission rates of over 10 Mbps are called ultra broadband systems. Broadband access systems that may be suitable for IPTV systems include

Chapter 6

digital subscriber line, cable modems, wireless broadband and powerline data.

Broadband access systems may be controlled (managed) or uncontrolled (unmanaged) systems. Broadband access systems that are controlled can guarantee the performance (data transmission rates). Broadband access systems that are uncontrolled offer best effort delivery.

If the broadband access system has data transmission rates that are several times the required data transmission rates for IPTV (2 to 4 Mbps), unmanaged systems may provide acceptable data transmission rates for quality IPTV programs.

Figure 6.4 shows the key types of broadband access providers that can be used to provide Internet television service. This diagram shows that some of the common types of broadband access systems that are available include

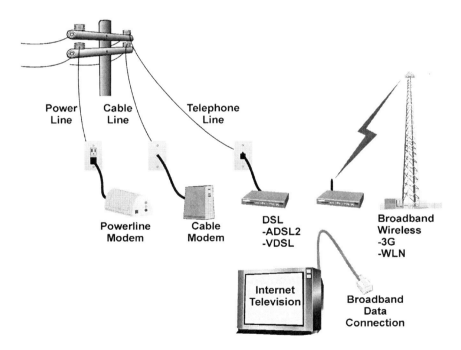

Figure 6.4, Broadband Access Providers

powerline data distribution, cable modems, digital subscriber lines (DSL) and wireless local area network systems (3G wireless, WLAN, MMDS, and LMDS).

Digital Subscriber Line (DSL)

Digital subscriber lines transmit high-speed digital information, usually on a copper wire pair. Although the transmitted information is in digital form, the transmission medium is an analog carrier signal (or the combination of many analog carrier signals) that is modulated by the digital information signal.

DSL systems have data transmission rates that range from 1 Mbps to over 52 Mpbs. DSL systems use dedicated wire connections between the systems digital subscriber line access modem (DSLAM) and the user's DSL modem.

The data transmission rates in DSL systems can be different in the downlink and uplink directions (asymmetric) or they can be the same in both directions (symmetric). Because IPTV data transmission is primarily from the system to the user (the downlink), asymmetric DSL systems (ADSL) are commonly used.

There are several types of DSL systems including asymmetric digital subscriber line (ADSL), symmetric digital subscriber line (SDSL) and very high bit rate digital subscriber line (VDSL). Of the different types of DSL, some have different versions with varying capabilities such as higher data transmission rates and longer transmission distances.

Because the telephone wires that DSL uses do not transfer high-frequency signals very well (high signal loss at higher frequencies), the maximum distance of DSL transmission is limited. In 2005, the typical maximum distance that DSL systems operated is 3 to 5 miles from the DSLAM. There is also a reduction in data transmission rate as the distance from the DSLAM increases. The further the distance the user is from the DSLAM, the lower the data transmission rate.

The data transmission required for each IP set top box in the home is approximately 2-4 Mbps. As the distance increases from the DSLAM to the customer, the number of set top boxes that can operate decreases. This means that customers who are located close to the DSLAM (may be located at the switching system) can have several set top boxes while IPTV customers who are located at longer distances from the DSLAM may only be able to have one IP set top box. As the demand for IPTV service increases, the DSL service provider can install DSLAMs at additional locations in their system.

Figure 6.5 shows how the number of simultaneous IPTV users per household geographic serving area can vary based on the data transmission capability of the service provider. This example shows that each single IPTV user typically requires 3 to 4 Mbps of data transfer. For a telephone system operator that uses distance sensitive DSL service, this example shows that the service provided will be limited to providing service to a single IPTV when

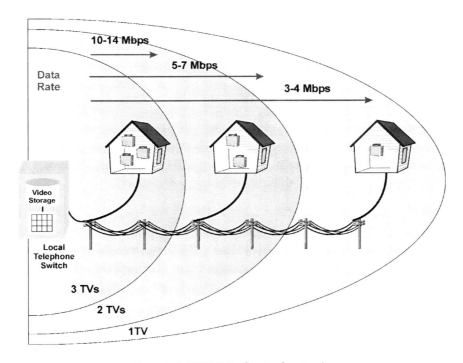

Figure 6.5, IPTV DSL Carrier Serving Area

the data transfer rates are limited to 3-4 Mbps. When the data transfer rate is above 5-7 Mbps, up to 2 IPTVs can be simultaneously used and when the data transmission is between 10 to 14 Mbps, up to 3 IPTVs can be simultaneously used.

Cable Modem

A cable modem is a communication device that modulates and demodulates (MoDem) data signals through a cable television system. Cable modems select and decode high data-rate signals on the cable television system (CATV) into digital signals that are designated for a specific user.

Cable modems operate over the same frequency band as a television channel (6 MHz for NTSC and 8 MHz for PAL). Converting a television channel to cable modem provides data transmission rates of 30-40 Mbps per television channel from the system to the user and 2-4 Mbps from the user to the cable system. Each coaxial cable is capable of carrying over 100 channels. Cable modem users on a cable television system typically share a data channel.

Cable television systems have traditionally been used to transfer analog television channels from the media source (cable company head end) to end users (televisions). To provide cable modem service, some of the channels (usually the higher frequency channels) are converted from analog signals to digital video and data signals. A single analog television channel can be converted to approximately 6 digital television channels.

DOCSIS is a standard used by cable television systems for providing Internet data services to users. The DOCSIS standard was developed primarily by equipment manufacturers and CATV operators. It details most aspects of data over cable networks including physical layer (modulation types and data rates), medium access control (MAC), services and security.

Figure 6.6 shows a basic cable modem system that consists of a head end (television receivers and cable modem system), distribution lines with amplifiers and cable modems that connect to customers' computers. This diagram shows that the cable television operator's head end system contains both analog and digital television channel transmitters that are connected to customers through the distribution lines. The distribution lines (fiber and/or coaxial cable) carry over 100 television RF channels. Some of the upper television RF channels are used for digital broadcast channels that transmit data to customers and the lower frequency channels are used to transmit digital information from the customer to the cable operator. Each of the upper digital channels can transfer 30 to 40 Mbps and each of the lower digital channels can transfer data at approximately 2 Mbps. The cable operator has replaced its one-way distribution amplifiers with precision (linear) high frequency bi-directional (two-way) amplifiers. Each high-speed Internet customer has a cable modem that can communicate with the cable modem termination system (CMTS) modem at the head end of the system where the CMTS system is connected to the Internet.

Wireless Broadband

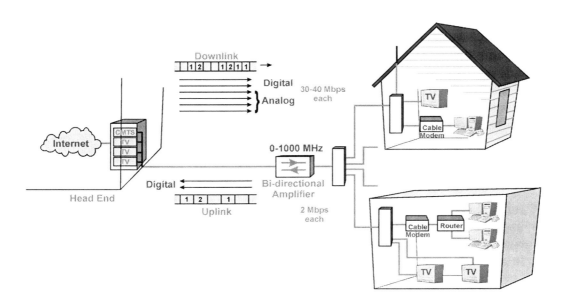

Figure 6.6, Cable Modem System

Wireless broadband is the transfer of high-speed data communications via a wireless connection. Wireless broadband often refers to data transmission rates of 1 Mbps or higher.

Some of the available systems that can offer wireless broadband services include satellite systems, fixed microwave (MMDS/LMDS), 3G mobile communication and wireless LAN. Wireless data transmission systems can transfer data rates from 1 Mbps to several hundred Mbps.

While the installation time for wireless systems is relatively short, the cost of wireless system bandwidth has been traditionally higher than wired systems. This is because wireless systems generally cannot multiply their bandwidth by adding more wired or lines. Another typical tradeoff for wireless system is higher mobility results in lower data transmission rates. Fixed wireless systems generally have higher data transmission rates than mobile wireless systems.

Satellite systems cover a very wide geographic area and they are well suited for one to many communication. The radio transmission from the digital broadcast satellite systems can provide data transmission at several hundred Mbps providing hundreds of digital video channels. While this system works well for simultaneous viewers, if the data transmission rate of each satellite channel is divided for number of viewers, the individual data rates will be relatively low (in the kbps). Some satellite systems use multiple beam transmission (spot beams), which can dramatically increase the available data transmission rates for each user.

Wireless broadband is commonly given the name "Wireless Cable" when it is used to deliver video, data and/or voice signals directly to end-users. Wireless cable provides video programming from a central location directly to homes via a small antenna that is mounted on the side of the house. There are two basic types of wireless cable systems, Multichannel Multipoint Distribution Service (MMDS) and Local Multichannel Distribution Service (LMDS).

MMDS is a wireless cable service that is used to provide a series of channel groups consisting of channels specifically allocated for wireless cable (the "commercial" channels). In the United States, MMDS service evolved from radio channels that were originally authorized for educational video distribution purposes. MMDS video broadcast systems have been in service since the early 1990's providing up to 33 channels of analog television over a frequency range from 2.1 to 2.69 GHz. Optionally, there are 31 "response channels" available near the upper end of the 2.5 to 2.69 band. These response channels were originally intended to transmit a voice channel from a classroom to a remote instructor.

MMDS systems normally transmit using an omni-directional antenna with the receivers, usually at a home, using a small directional antenna. The service provider that converts the MMDS frequencies to ordinary VHF TV channels supplies a frequency down-converter.

Local multipoint distribution service is a wireless broadband distribution system that operates in the 28 GHz to 31 GHz frequency band. In the United States, LMDS entered into the FCC auction process in 1997.

LMDS uses approximately 1.3 GHz wide spectrum band near 28 GHz. This provides a typical data rate for each LMDS channel of 1 Gbps. Because of the extremely high frequencies used, the transmitter must be located within 3 to 5 miles of the receiver. The limitation of short distance is that LMDS signals from one antenna will not interfere with other antennas placed 10 or more miles apart. This allows the radio bandwidth to be reused (frequency reused) in a cellular like fashion.

Initially, radio coverage was perceived to be a challenging factor for LMDS service as 28-31 GHz microwave signals cannot regularly penetrate into buildings very well. This would have limited the deployment of LMDS systems into large cities such as New York City. However, because microwave signals reflect off large objects (such as buildings), it is possible to exploit the reflections off buildings to provide extended coverage.

The key challenge for LMDS operators will likely be the initial construction of systems. Unlike cellular telephone operators, which were able to start with large cells and gradually split them into smaller cells, LMDS systems can only offer small radio coverage areas (cells). This means that LMDS service providers will likely target areas with a large concentration of potential users.

LMDS frequencies in the United States are near 28 and 29 GHz downlink and near 31 GHz for an uplink return path (transmission from the end customers equipment to the system operator).

Multipoint video distribution system is a wireless broadband distribution system that operates in the 40.5 GHz to 43.5 GHz frequency band. The MVDS are defined by the European Radiocommunications Committee (ERC) in 1999.

In 1996, the ERC designed the frequency band of 40.5 GHz to 42.5 GHz for the distribution of television programming. In 1998, the ERC took into account the requirements of multimedia wireless systems (MWS) and expanded the frequency band from 42.5 GHz to 43.5 GHz and required all future systems in these bands use digital radio transmission. While the system is primarily designated for broadcast, it does allow some interactivity.

Figure 6.7 shows that the key components of a wireless cable system include the head-end equipment and end user equipment. The head-end equipment is equivalent to a telephone central office. The head-end building has a satellite connection for cable channels and video players for video on demand. The head end is linked to base stations (BS), which transmits radio frequency signals for reception. An antenna and receiver in the home converts the microwave radio signals into the standard television channels for use in the home. As in traditional cable systems, a set-top box decodes the signal for input to the television. Low frequency MMDS wireless cable systems can reach up to approximately 35 miles. LMDS systems can only reach approximately 3-5 miles.

The 3rd generation wireless requirements are defined in the International

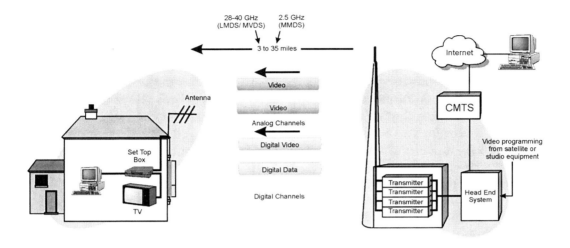

Figure 6.7, Wireless Cable System

Mobile Telecommunications "IMT-2000" project developed by the International Telecommunication Union (ITU). The IMT-2000 project that defined requirements for high-speed data transmission, Internet Protocol (IP)-based services, global roaming and multimedia communications. After many communication proposals were reviewed, two global systems are emerging; wideband code division multiple access (WCDMA) and CDMA2000/EVDO.

The maximum data transmission rates for 3G mobile communication systems can be over to 2 Mbps. However, the highest data transmission rates only can be achieved when the user is located close to the radio tower (cell site) and their interference levels (signals from other towers and users) are low. As a result, much of the live digital video transmission for mobile communication systems in 2005 had data transmission rates of approximately 144 kbps.

A wireless local area network (WLAN) is a wireless communication system that allows computers and workstations to communicate data with each other using radio waves as the transmission medium. The 802.11 industry standard and its various revisions are a particular form of Wireless LAN.

802.11 WLAN is commonly referred to as "Wi-Fi" (wireless fidelity). To help ensure Wi-Fi products perform correctly and are interoperable with each other, the Wi-Fi Alliance was created in 1999. The Wi-Fi Alliance is a non-profit organization that certifies products conform to the industry specification and interoperates with each other. Wi-Fi® is a registered trademark of the Wi-Fi Alliance and the indication that the product is Wi-Fi Certified™ indicates products have been tested and should be interoperable with other products regardless of who manufactured the product.

Wireless LANs can be connected to a wired LAN as an extension to the system or it can be operated independently to provide the data connections between all the computers within a temporary ("ad-hoc") network. WLANs can be used in both indoor and outdoor environments.

Wireless LANs can provide almost all the functionality and high data-transmission rates offered by wired LANs, but without the physical constraints of the wire itself. Wireless LAN configurations range from temporary independent connections between two computers to managed data communication networks that interconnect to other data networks (such as the Internet). Data rates for WLAN systems typically vary from 1 Mbps to more than 50 Mbps.

Wireless LAN systems may be used to provide service to visiting users in specific areas (called "hot spots"). Hot spots are geographic regions or service access points that have a higher than average amount of usage. Examples of hot spots include wireless LAN (WLAN) access points in coffee shops, airports, and hotels.

While 802.11 WLAN systems are well suited for premises distribution, because of its limited range of approximately 100 meters, 802.11 Wi-Fi systems are not likely to be the preferred broadband access connection.

WiMAX is a name for the IEEE 802.16A point to multipoint broadband wireless industry specification that provide up to 150 Mbps data transmission rate in either direction with a maximum transmission range of approximately 3-10 km [1]. The WiMAX system is primarily defined to operate in either the 10 GHz to 66 GHz bands or the 2 GHz to 11 GHz bands.

Like the short range 802.11 wireless local area network (WLAN) specification, the 802.16 systems primarily defines the physical and media access control (MAC) layers. The system was designed to integrate well with other protocols including ATM, Ethernet, and Internet Protocol (IP), and it has the capability to differentiate and provide differing levels of quality of service (QoS).

Power Line Carrier (PLC)

Power line carrier is the simultaneous sending of a carrier wave information signal (typically a data signal) on electrical power lines to transfer information. A power line carrier signal is above the standard 60 Hz or 50 Hz frequency that transfers the electrical energy.

Power line carrier systems used in wide area distribution transfer signals from a data node located in the power system grid to a data termination box located in the home. Power line carrier systems typically use multiple transmission channels (low and medium frequencies) to communicate between the power system data node and the data termination box.

Because power lines were not designed for low and medium frequency signals, some of the signal energy leak from the power lines and some undesired signals are received into the power lines from outside sources (such as AM radio stations). The interference is typically different for each frequency so each power line transmission carrier signal is handled independently. If the signal energy is lost or if other signals interfere with one frequency, one or several transmission channel frequencies can be unused. Multiple transmission channels can be combined to provide high-speed data transmission rates.

Powerline transformers typically block high frequency data signals so data routers or bridges are typically installed near transformers. Utility companies typically setup data nodes near transformers and each data node provides signals to homes that are connected to the transformers. There are often many homes per transformer (over 1,000 homes) in developing countries and a lesser number of homes per transformer (less then 12 homes) in developed countries.

Figure 6.8 shows three types of communication systems that use an electrical power distribution system to simultaneously carry information signals along with electrical power signals. In this example, the high voltage portion of the transmission system is modified to include communication transceivers that can withstand the high voltages while coupling/transferring their information to other receivers that are connected to the high voltage lines. This type of communication could be used to monitor and control

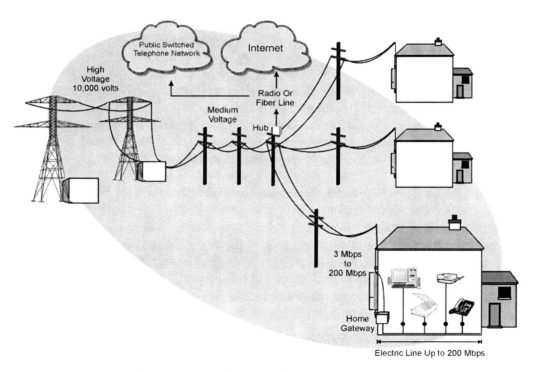

Figure 6.8, Powerline Broadband Data Transmission

power distribution equipment such as relays and transformers or as a high-speed data backbone transmission connection using fiber. This example also shows a power line distribution system that locates a communication node (radio or fiber hub) near a transformer and provides a data signal to homes connected to a transformer. This system could allow customers to obtain Internet access or digital telephone service by plugging the computer or special telephone into a standard power socket. The diagram also shows how a consumer may use the electrical wiring in their home as a distribution system for data (e.g. Ethernet) communication.

Headend

The headend is the master distribution center of an IPTV system where incoming television signals from video sources (e.g., DBS satellites, local studios, video players) are received, amplified, and re-modulated onto TV channels for transmission into the IPTV distribution system.

The incoming signals for headend systems include satellite receivers, off-air receivers and other types of transmission links. The signals are received (selected) and processed using channel decoders. Headends commonly use integrated receiver devices that combine multiple receiver, decoding and decryption functions into one assembly. After the headend receives, separates and converts incoming signals into new formats, the signals are selected and encoded so they can be retransmitted (or stored) in the IPTV network. These signals are encoded, converted into packets and sent into IPTV packet distribution system.

Figure 6.9 shows a diagram of a simple headend system. This diagram shows that the headend gathers programming sources, decodes, selects and retransmits video programming to the IP distribution network. The video sources to the headend typically include satellite signals, off air receivers, microwave connections and other video feed signals. The video sources are scrambled to prevent unauthorized viewing before being sent to the cable distribution system. The headend receives, decodes and decrypts these channels. This example shows that the programs that will be broadcasted are supplied to encoders produce IP television program streams. The pro-

IPTV Business Opportunities

grams are sent into the IPTV distribution network to distribution points (e.g. media servers) or directly to end users devices (e.g. set top boxes).

An IPTV system has expanded to include multiple regions including local

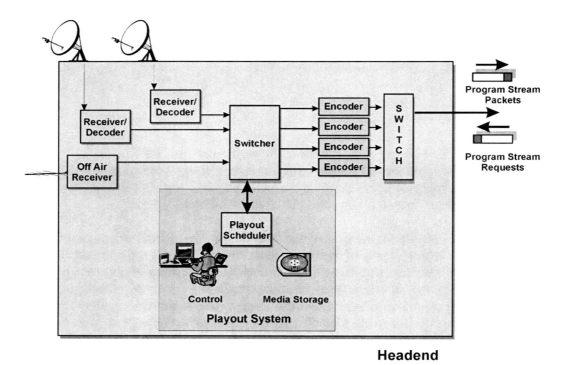

Figure 6.9, Headend System

headend locations, which may be distributed over a large geographic region. Local headends may be connected to regional headends and regional headends may be connected to a super headend. To reduce the cost of an IPTV system, headend systems can be shared by several distribution systems.

Chapter 6

Contribution Network

A contribution network is a system that interconnects contribution sources (media programs) to a content user (e.g. a television system). IPTV systems receive content from multiple sources through connections that range from dedicated high-speed fiber optic connections to the delivery of stored media. Content sources include program networks, content aggregators and a variety of other government, education and public sources.

In additional to gathering content through communication links, content may be gathered through the use of stored media. Examples of stored media include magnetic tapes (VHS or Beta) and optical disks (CD or DVDs).

When content is delivered through the content network, its descriptive information (metadata) is also delivered. The metadata information may be embedded within the media file(s) or it may be sent as separate data files. Some of the descriptive data may include text that is used for closed captioning compliance.

Figure 6.10 shows a contribution network that is used with an IPTV system. This example shows that programming that is gathered through a contribution network can come from a variety of sources that include satellite connections, leased lines, virtual networks, microwave links, mobile data, public data networks (e.g. Internet) and the use of stored media (tapes and DVDs).

IPTV Business Opportunities

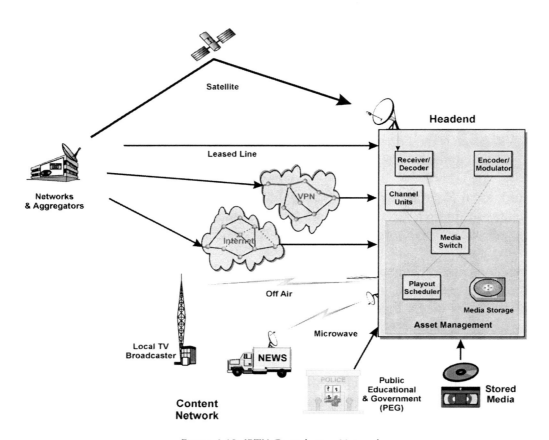

Figure 6.10, IPTV Contribution Network

References

[1]. "What is Wi-Max?", WiMAX Forum, www. WiMAXforum.org.

Chapter 7

IPTV System Types

IPTV system types include managed IPTV, unmanaged IPTV (Internet TV) and private IPTV systems.

Managed IPTV systems provide services to customers using controlled access Internet protocol (IP) connections. Managed IPTV systems allow customers to have the quality and features that are typically associated with a traditional television systems (such as cable television – CATV). Because the IPTV provider manages (controls) the bandwidth of the access network, it is possible to guarantee the video quality and reliability of the IPTV service to the customer.

Managed IPTV systems include IPTV over cable, telephone ("TelcoTV"), IPTV over cable modem, wireless broadband or digital television over powerline where the IPTV provider controls both the access network and the services.

Unmanaged IPTV (also called Broadband television) is the delivery of digital television services over uncontrolled broadband data networks. It may be possible to control and guarantee the quality of television services if the underlying broadband connections have enough bandwidth. Internet service providers or media management companies usually provide unmanaged IPTV systems through broadband Internet connections.

Internet television service providers (ITVSPs) supply IPTV services to their customers that are connected to broadband Internet connections. ITVSPs provide access screens or channels links allow the customer to connect their viewers to media sources for a fee. ITVSPs can vary from companies that simply provide IPTV links to viewers to media content providers.

Customers use IPTV access devices (media players) or IP set top boxes (self contained media players) to communicate their requests for services and features to the ITVSP. The ITVSP receives their requests and determines what features they are authorized to access that particular feature or service. If the ITVSP decides to provide service, it will determine which gateway it will use to connect the viewer to the media source. The gateway will record the time and usage information and send it to the ITVSP to account for the usage (to get paid).

ITVSP companies are primarily made of computers that are connected to the Internet and software to operate call processing and other services. ITVSPs use computer servers to keep track of which customers are active (registration) and what features and services are authorized. When television channel requests are processed, the ITVSP sends messages to gateways via the Internet allowing television channel to be connected to the viewer's media player.

Because ITVSPs provide IPTV signals through public broadband networks, they cannot directly control the quality of service for data transmission. While the ITVSP cannot directly control or guarantee IPTV quality, if the data links in the connection from the media source to the viewer have data rates that are much higher than the viewer's data rate, the channel quality is likely to be good for the selected digital video signal.

Chapter 7

Internet television service providers (ITVSPs) help customers find Internet television channels and manage connections between media sources. While it is possible in some cases for end users to directly connect to media source by using an IP address or even a web link, Internet television provides may simplify the programming guide choices. Internet television service providers may also provide connections to subscription controlled television sources. For this role, the IPTV service provider makes a business relationship with the media source. The IPTV service provider may pay the media provider from funds it connections from their end user's.

Figure 7.1 shows that Internet television service providers (ITVSPs) are primarily made of computers that are connected to the Internet and software to operate call processing and other services. In this diagram, a computer keeps track of which customers are active (registration) and what features and services are authorized. When television channel requests are processed, the ITVSP sends messages to gateways via the Internet allowing

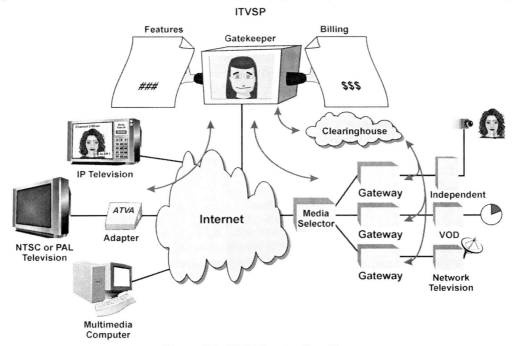

Figure 7.1, IPTV Service Providers

133

television channels to be connected to selected media gateway source (such as television channels). These media gateways transfer their billing details to a clearinghouse so the ITVSP can pay for the gateway's usage. The ITVSP then can use this billing information to charge the customer for channels viewed.

On Network (On-Net)

On network is the providing or controlling of services within the network that the operator owners, manages or controls. An example of "on network" is the transmission of a television program from a media server through a data network to a set top box that the IPTV system operator directly controls or manages.

Off-Network (Off-Net)

Off network is the providing or controlling of services within networks that other companies own, manage or control. An example of an "off network" service is the providing of access to a television program through a set top box that is controlled by an IPTV operator to an international television program provider where the content is delivered through the Internet which the IPTV system operator does not control.

Over The Top

Over the top services are applications and their associated communication transmission provided for the benefit of the user through the use of underlying services. An example of an over the top service is a telephone or television service that is provided over an Internet connection.

Chapter 7

Hybrid IPTV Systems

Hybrid IPTV systems use two or more system types to provide IPTV services. Hybrid IPTV systems may be used to bridge the gap between managed IPTV (TV over telephone lines) and unmanaged broadcast services (broadband TV). These systems may provide the customer with a mix of high quality content sources (HD and SD) along with high value (HV) content (such as international programs distributed through Internet connections).

System Operation Options

IPTV system operators may control all portions of their network (network operator), some of their network (shared network) or resell services through other networks and systems (virtual operator).

Network Operator

An IPTV network operator is a company that controls and manages both the access network and IPTV system equipment. IPTV network operators have the advantage of having full control of the network, which allows them to rapidly trial and offer new features and services.

Operational costs for network operators can be high due to the necessity to have people with skill sets ranging from packet switching to content licensing. This is especially true for small to medium size IPTV system operators where the costs of maintaining staff with a mix of skill sets is divided over a small number of customers. Small to medium size operators can make alliances with other nearby operators to share staff with the necessary technical skill sets when required.

Small to medium size network operators that have a limited number of customers will purchase content in smaller quantities than other types of media providers. This means that the cost of content may be higher.

The initial investment per subscriber for shared networks can range from $1,000 to $2,000 because the shared network operator must purchase and install access distribution equipment and IPTV media equipment and software. Because IPTV network operators must setup both content acquisition and distribution systems, time to market can be 1 to 2 years.

Shared Network

Shared networks are systems that allow companies to share and control the resources of other networks or systems. An example of a shared IPTV network is an IPTV operator that partners with a broadband access network (such as a WiMAX operator).

Shared networks is an Off-Net (Off the Controlled Network) solution where control and management of systems and services may be limited. Shared network operators may have remote control of access and full control of content network operation. Shared network operators may have control of the content selection, headend, distribution and access network through the use of interface controls between networks. While shared network operators may be able to control and provision systems for quality of service, shared network operators may have limited controls or responsibilities that can be difficult to define (not my fault). This may lead to challenges resolving QoS issues.

A key advantage of using a shared network is that operators can expand into geographic areas where they can find partner networks. Because the cost of the network is shared by two or more partners, the key objective of the shared network operator is to focus on their part such as programming and customer care.

Operational costs can be moderate for shared network operators. The number of skill sets need to operate and manage the content and distribution network is reduced as each partner has their own support systems.

Because shared network operators may be able to control and manage their networks, these systems can have strong security controls. If the shared network operator can demonstrate that their system is secure, content sources are likely to be willing to let the network operator to distribute content if the right digital rights management systems are in place. Because shared network operators may have several systems with many customers, the cost of content may be lower because of the increased volume of content demand.

The initial investment per subscriber for shared networks can range from $200 to $500 because the shared network operator does not need to purchase and install access distribution equipment. Because the access network is already setup, to market that is necessary to setup shared systems can range from 6 to 12 months.

Virtual (Hosted) Network

A virtual network operator is a service provider that uses the communication systems of other communication network operators to provide communication services. Virtual operators can be resellers of managed IPTV systems or they can be are over the top services (Broadband TV).

Virtual operators primarily control of the selection of media that will be offered to their customers. Virtual operators need to focus on learning the restrictions for content distribution and delivery in the area they serve (possibly in several countries). Virtual operators use systems, services and protocols (control processes) that are developed by other companies, which may limit their ability to customize the services they offer. While the control of features and services may be set by the virtual host, virtual operators may be able to customize features and services through the use of programming scripts or software applications

Operational costs can be low for virtual operators, even for smaller systems. The number of skill sets need to operate and manage the content and distribution network can be a very small number of people. The operational costs for virtual operator's shifts from running a network to managing content that is promoted and distributed through the network partners.

The content offered by virtual operators can range from a limited number of specialty channels to popular network programming. Because virtual operators may not be able to control all the security components of programs transmitted through an underlying network, it may be more difficult to obtain licenses to distribute more popular content.

The capital cost (Capex) per subscriber for virtual operators can range between $0 to $200. The key capital cost is the IPTV STB, which is approximately $100 to $200. Customers may be required to buy their own STB. Because the customer interest is to obtain programming that may not be available anywhere else, this gives the virtual operator added ability to charge setup and equipment fees.

The time required for a virtual operator to get to market can be weeks to months because the systems are already setup and only the content, branding and distribution channels need to be established.

Figure 7.2 shows several options that Internet IPTV operators have to provide television services. This table shows that IPTV system operators can own and operate their own system, they can run a system that is owned by another company or they can use an Internet IPTV host company.

Network Operator	Shared Network	Virtual Network
Full control of access network and content network	Remote control of access network and full control of content network.	Control of Features and Services
Operational costs high due to multiple required skill sets (network + content)	Operation cost mid-range due to shared costs (network or content)	Operational cost low due skill set requirements (content only)
Content cost high for network programming	Content cost reduced due to higher number of viewers	Content cost can be low for non-network (specialized) programming
Capital Cost per Subscriber $1,000 to $2,000	Capital Cost per Subscriber $200 to $500	Capital Cost per Subscriber $0 to $200
Time to market can be 1 to 2 years	Time to market can be 6 to 12 months	Time to market can be weeks to months

Figure 7.2, IPTV System Options

Chapter 7

Private IPTV Systems

Private IPTV systems are used to provide television service within a building or group of buildings in a small geographic area. Private IPTV systems typically contain media servers, media gateways and advanced television channel selection processing features such as channel selection and media control (play, stop, fast forward). Applications for private IPTV systems include hotel television, community television, surveillance security and video conferencing. Private IPTV systems may be directly connected to media sources (e.g. DVD video players) and/or they may be connected to other media sources through gateways.

Private IPTV services include television, on demand media, high speed Internet access, interactive applications and telephony. Private IPTV systems commonly use a rugged version of an IP set top box with some added features. If a company that is installing a private IPTV system already has an existing television system (such as a hotel TV system), they may implement a hybrid version of IPTV systems. Hybrid IPTV systems use existing RF equipment and cabling for the transmission of popular network channels while using IP data networks to delivery IPTV on-demand programming [1].

There are several ways that private television systems (such as CATV systems) can be upgraded for IPTV capability. The options range from adding IPTV gateways as a media source to allow the existing television systems to receive new television channels to replacing the media sources and set top boxes with IP video devices.

Private IPTV system owners may obtain the content for the TV systems from a combination of public television (such as cable TV and off-air) and alternative content sources (such as international channels and adult content providers). Private IPTV system owner may also obtain revenue from ad insertion into channels where they control the content (such as ads on private television channels in gyms, restaurants and hotels).

Private IPTV Benefits

Some of the key benefits of using IPTV for private systems (such as in hotels) include reduced television network equipment cost, simplified operation and new revenue producing services as compared to other RF television systems.

Private IPTV systems use standard Internet protocol data networks to setup, deliver and manage video services. Because IP data communication equipment is available from many vendors, the cost of standard IP datacom and supporting equipment can be much lower to purchase, install and maintain than customized RF systems.

IPTV television programming sources usually includes the standard network program channels. Fully configured private IPTV systems have the potential to offer thousands of channels (or more), which means that private IPTV systems can offer many international and specialty programming choices.

Because each IPTV STB can select and play programs through its own data connection to media sources, this means that users to select, watch and control the playing of movies at any time. Private IPTV systems can offer real video on demand services providing instant access to more programs.

IPTV systems use digital media that can be stored and access from reliable sources such as memory storage, hard disk drives or DVDs. The use of all digital media provides more standard and improved digital quality. Using IPTV systems, it is also possible to provide and unlimited number of high definition (HD) channels. The use of digital storage systems can eliminate the use of mechanical systems such as video cassette records (VCRs) which will increase the reliability of the system.

The ability to direct connect private IPTV systems through Internet data connections permits for the remote configuration and control. This enables for remote monitoring and remote diagnostics to rapidly repair failed or poor performing equipment. The remote monitoring also permits revenue assurance which corporations and content providers desire to ensure their content is paid for when actually used.

Content for private IPTV systems can be distributed through the same Internet data connections used by the owner of the IPTV system (such as a hotel's Internet connection). By using encryption (data scrambling), content can be securely transferred from content providers into the private IPTV system. This allows for more rapid and cost effective distribution of content.

IPTV systems use a single type of data connection for television, data and telephony services. This allows the IPTV system to use a single set of cabling instead of separate telephone, data and television wiring systems. Because of the ability of IP data systems to automatically reconfigure, even the single set of data cables can be very reliable.

Figure 7.3 shows some the key benefits for deploying IPTV systems. This table shows that private IPTV systems have many benefits that range from standard and shared data distribution systems to the potential ability to offer instant selection and control of an unlimited number of programs.

Capability	Benefit
IP Distribution	Can use low cost IP data equipment
Unlimited Channel Selection	Offer many more television programs
Independent Channel Control	Real (anytime) on demand program access and control
Digital Media Format	Better quality video and enables the use of more reliable media storage such as memory or DVD
Remote Access Control	Remote monitoring, diagnostics and updating
Download Content Distribution	Secure transfer of programs
Shared Infrastructure	Single set of cabling that is shared by television, data and telephone

Figure 7.3, Private IPTV Benefits

Private IPTV Equipment

The equipment needed for private IPTV systems includes ruggedized IP set top boxes, IP data distribution systems and a media head end. Private IP set top boxes are adapters that convert IP data into television video signals. Private STBs may also be used to connect other types of devices that are located in the room where they are located (such as in a hotel room). These connections can include data ports (for Internet connections), telephone jacks and other accessories (such as game controllers and video cameras). The STB becomes the portal that allows other devices such as game controllers, computer data connections and telephones to be connected through a single data communication system.

Set top boxes used in the private IPTV systems may need to be ruggedized and difficult to steal. IP STBs used in hotels or schools may be bolted underneath or behind furniture to make them less accessible to users. To allow operation with a remote control that requires line of sight, hospitality STBs may use a remote infrared receiver. The remote receiver is mounted near the television and is connected by wire to the hospitality STB. When the STB also provides an Internet for the user, it can reduce the bandwidth so they can limit the quality of the digital video signal to the room (and so they can't watch video directly through Internet).

Private TV Content Sources

One of the key issues for private IPTV systems is where to get content. Content distributors commonly charge hotels a daily fee for day per room regardless if it is occupied or not. Popular content owners (such as movies) are protective about where, who and how users view content.

IPTV systems can accurately track and record exactly who (or what room) watched what programs and when they were watched. This is a great way to provide revenue assurance for content providers and to control costs for unused rooms.

Private IPTV systems commonly receive network programs from either cable television or satellite connections. They also may connect receivers to antennas to get free local broadcast (off air signals) channels.

In addition to traditional programming sources, new content providers can stream or download new channels through the Internet that can be made available to the IPTV system. These include international programs (such as international sports events) and specialty channels. On demand entertainment, especially adult entertainment, offers relatively high margins and it can be a big part of hotel's television service revenues.

Another emerging service that private systems can offer is local information channels and advertising. While it is not clear on how advertising messages can be inserted, there is significant potential for capturing revenue for inserted ads.

References:

[1]. "Hosptiality IPTV", IPTV Magazine, www.IPTVMagazine.com

IPTV Business Opportunities

Chapter 8

IPTV Economics

The economic goal of an IPTV system is to effectively provide revenue producing services to many customers at the lowest possible cost. The ability to serve customers is determined by the capacity of the IPTV system and its distribution network along with the capabilities of the software (middleware) the system uses.

Revenue

Revenue services for IPTV system operators include video services, data services, voice services, advertising and television commerce. Video services are the largest revenue producing service for cable operators.

Because IPTV system operators provide multiple (bundled) services, the average revenue per user can be relatively high. For example, the average revenue per user (ARPU) for Time Warner's cable service which is composed of video, data and voice services was approximately $95 in 2007 [1].

Revenue increases for television service providers have come from increases in service fees and the sale of additional services (such as subscription fees for additional premium video channels or data service fees). As more media distributor options become available, increases in video service rates may be less effective or not possible. Video services can be divided into basic, expanded and premium services.

Basic Services

Basic video services account for approximately ½ of service revenue. Basic services include a mix of local content and network programming.

Extended Services

Extended services contribute approximately 15% of the service revenue. Extended video services include specialty program groups such as children's and entertainment packages.

Premium Services

Premium services generate near 12% of service revenue. Premium services include movie channels and pay per view programming. Due to the ability to offer virtually unlimited on demand programming for IPTV systems, the revenue for premium services is likely to increase.

Internet Access

Internet access (data services) is responsible for approximately 9% of service revues. The Internet access trend is an increasing demand for higher-speed data services to provide access to advanced media services (such as video clips).

Voice Services

Voice service revenue contributes 4% to 6% of the subscriber revenue. In general, there is a decline in the service plan rates for voice services without a corresponding increase in consumption. The trends for providing voice service include unlimited regional calling plans increasing geographic territory for calling plans.

Advertising

Advertising revenue for television providers contributes approximately 4% of revenue. Advertising revenue varies based on viewer ratings. A key growth for television advertising is local targeted advertising. IPTV systems are ideally suited to deliver local advertising.

IPTV service providers have limited amounts of advertising rights for traditional programming channels. For specialty programming (such as international channels), IPTV service providers may have rights to insert ads. Internet advertisers (such as Google) are entering into the television advertising business. This will allow for the dynamic insertion of ads to viewers and providing an automatic system for selling ads for IPTV service providers.

IPTV service providers may receive commissions for the sale of items on shopping channels. This may be recorded as advertising income, but it actually is television commerce revenue.

Because of the ability of IPTV systems to provide highly targeted advertising, the revenue contribution from advertising is likely to dramatically increase.

Digital Services

The contribution of digital services such as web page hosting, personal media management and backup storage systems is approximately 2% of service revenue.

Television Commerce (T-Commerce)

Television commerce is an emerging category for IPTV service providers. While we could not find a category called television commerce on financial statements, T-Commerce offers the highest potential revenue source for IPTV service providers. In 2008, close to $1 trillion is spent by consumers on e-commerce worldwide [2]. Television commerce (t-commerce) is a form of e-commerce.

As IPTV systems evolve, they will be better suited to provide t-commerce as most viewers spend more time in front of the television than in front of a computer each day and when they are watching television, they are more relaxed and receptive to promotional media.

Other

Other revenue sources include affiliate content providers, equipment sales and rental, installation and connections, along with other unclassified revenue sources. These revenue sources are relatively small (approximately 2%).

Figure 8.1 shows an approximate split of revenue for an IPTV service provider in 2008. This table shows that a majority of the revenue is created from video (basic, extended and premium content) services. Internet access (9%) and voice services (6%) are important parts of revenue. Advertising is a small part of revenue as it can only be sold and inserted in a limited number of program channels. Digital services such as web hosting and other services are a small part. The television commerce (t-commerce) category has not yet been separately categorized and developed by most IPTV carriers.

Revenue Category	Revenue Percentage
Basic services	50%
Extended services	15%
Premium services	12%
Internet access	9%
Voice Services	6%
Advertising	4%
Digital services	2%
Other	2%
Television Commerce	0%

Figure 8.1, Sample IPTV Revenue Sources 2008

Costs

IPTV costs are approximately 70% of revenue providing a net margin of approximately 30%. The leading cost for IPTV systems is content cost. Other IPTV costs include operations, general administration, marketing, voice transmission cost and data transmission costs.

Content Costs

Content cost for an IPTV service providers ranges from approximately 40% to 50% of overall expenses. IPTV operators may pay more for content as the amount of content they provide is a small percentage compared to larger broadcasters. For example, the average video content cost per user for Time Warner's cable service was approximately 31% of revenue from video services ($21 per month) [3].

Obtaining content can involve multi-year licensing agreements for video programming and these agreements can have minimum license fee guarantees. Many financial reports for cable operators in 2008 indicated they expected the cost for programming content to increase due to contractual requirements.

Over the past few years, the average cost of video programming to television subscribers has continued to increase. To help attract subscribers to IPTV service, some IPTV carriers have offered additional content programming sources that have a mix of content costs. Currently, much of these alternative content costs are low. However, as the demand for alternative programming increases as new IPTV system operators emerge, the cost of quality alternative programming is likely to increase. However, the cost of alternative programming will likely to be lower than traditional video content costs (network programming).

When IPTV system operators begin to offer service, the number of viewers they have is small, which means the IPTV service provider typically pays more (lower volume discount) for traditional programming channels than cable TV and satellite systems operators pay.

As a result of the high cost of traditional programming, IPTV service providers have been searching for alternative high-value program sources. Some alternative content solutions for IPTV providers include content owners paying to create content and sponsored solution content.

Content owners such as governments, schools and religious groups have started to pay companies such as New Century Television (www.NewCenturyTV.com) to produce linear television programming and to host this programming on broadband networks. Sponsored solution content is specialty programs that are paid for by companies who are looking to promote a specific solution or to develop a mailing list of the viewers who watch the solution content (similar to sponsored Webinars).

Operations

The costs of operating an IPTV system include the costs of installing, maintaining and repairing communication systems. Because of the need for a mix of technical skills to maintain IPTV systems, the approximate cost of operations for IPTV systems is 20%, which may be higher than cable companies (Time Warner cost of operation staff cost in 2007 was approximately 13% [4]).

General Administration

General administration costs are facilities and staff costs that are required to provide basic support services (such as accounting and human resources). General administrative costs for IPTV systems approximately 12% of cost. General administration cost may include management fees of 5% to 10% for smaller to mid size companies that rely on other companies to perform management services. This is higher than general administration cost for large cable companies such as Time Warner that has approximately 6% general administration cost in 2007 [5].

Marketing

Marketing costs for television broadcasters and cable can be approximately 2% to 4% of revenue (Time Warner marketing was approximately 3% of revenue in 2007 [6]). The key marketing factors that may determine the success of IPTV systems includes the types of content, new services, new features, consumer confidence and pricing of combined communication service.

Traditional television marketing programs have focused on upselling existing customers to purchase addition programming packages and to add related services such as high-speed data and telephone services.

The television industry is undergoing a dramatic change. New television service providers are emerging from other industries that will compete in the marketplace with new competitive features. This is likely to increase competition in the television industry. As a result, sales and distribution channels may become clogged with a variety of television product offerings.

To better complete, existing television providers (cable and satellite) may shift their focus to subscriber retention rather than on upselling. Similar to the mobile communication industry, these tactics may include subscriber contracts and early disconnect penalties. This is likely to make it harder to capture customers from competitors and increase marketing costs.

Data Services Costs

Data services costs include data storage services, leased lines and Internet connectivity costs that are associated with providing data services (such as Internet connectivity). Data service costs are approximately 2% to 4% of average customer service revenue (15% to 25% of the revenue for data service).

Internet connection costs are based on data transmission speed and amount of data transferred over a period of time. Because media content on the Internet is evolving to rich media (photos and video), data services costs are likely to increase as consumers have access to and consume to more rich content.

Voice Service Costs

Voice service costs include network equipment to provide voice services, leased line cost and telephone network connectivity costs. The cost of providing voice services can be approximately 4% of service revenues.

The data transmission bandwidth needed for voice services is very low as compared to the data transmission bandwidth for video or other multimedia services. The primary cost for providing voice communication services is the interconnection and delivery cost through other telephone networks.

Since the mid-1990s, there have been changes in interconnection costs to telephone companies. With the deregulation of telecommunications systems, companies have resulted in service reciprocity requirements. Reciprocal compensation is the process of accessing similar costs for similar services. The 1996 Telecommunications Act mandated that local telecommunications companies exchange revenue for the cost of terminating calls that originated on the wireline network. Previously, only wireless companies were obligated to pay compensation for calls originated on their networks but terminated on the wireline network. As a result of reciprocity requirements, the cost of interconnection has declined from approximately 4 cents per minute to less than 1 cent per minute.

Figure 8.2 shows an approximate split of costs for an IPTV service provider in 2008. This table shows that a majority of the costs come from video content (45%). The costs associated with IPTV system operation is 20% and general administrative and overhead costs are 12%. Marketing costs are 6%. The cost associated with voice services (telephone system interconnection) are approximately 4% and data services (Internet connection) are 2%. Other non-categorized costs total is 11% of the cost to provide services.

Expense Category	Expense Percentage
Programming Costs	45%
Operations Cost	20%
General Admin	12%
Marketing	6%
Voice Services	4%
Data Services	2%
Other	11%

Figure 8.2, Sample IPTV Cost Sources 2008

Capital Costs

IPTV capital costs are long term expenses that include the purchase of land, buildings and most importantly, the build out of network capacity in communication system. Capital cost per subscriber is the total cost of network investment required to provide services to subscribers (customers).

The initial cost (capital cost) of adding IPTV capability to existing distribution systems (such as telephone systems) can range from approximately $400 to $700 per subscriber for systems ranging from 2,500 to 10,000 video subscribers. [7].

Options for implementing IPTV systems include purchasing (building) a headend system or choosing a service provider headend solution. Purchasing a digital headend includes acquiring and installing equipment that can receive national channels, gathering and retransmitting local off air channels and connecting and distributing other video or media sources. Deploying a shared headend service also requires upfront purchasing of equipment. Both options typically require purchasing and installing at least one or more satellite dish and receivers. The exception to requiring a satellite receiver dish is when a Telco receives its main channels from a cooperative that can be using fiber based networks to haul the signals to the remote Telco members.

A key difference between the dedicated headend option and the shared service provider headend option is the percentage of revenue that is paid for content. The dedicated headend system has a content cost of approximately 50% where the service provider headend system has higher content cost up to approximately 75% of revenue per subscriber.

Headend

The headend is composed of receivers, channel decoders, channel encoders and network transmission equipment. Headends may also contain additional equipment such as switching and video production equipment and facilities (studios).

The headend is usually located in a long-term location (10-20 years), usually near a high-speed data network connection point. Commonly, the customer databases are located in the headend facility. The headend system commonly contains basic software that allows normal television service operation (distribute and process television connections). Additional features often require software applications that allow advanced services are available at additional cost.

For the shared headend systems, fiber or satellite feeds from a shared distribution may be used instead of purchasing encoders and other headend media equipment. The use of shared headend systems can reduce the initial capital equipment cost by 70% or more.

Middleware

Middleware is a software system that links the headend equipment to the access devices (e.g. set top boxes). A middleware system has a one time acquisition and setup cost of approximately $150,000 and a one time per STB subscriber licensing fee of $25.

Conditional Access and DRM System

Conditional access and digital rights management (CA/DRM) systems are needed, which cost approximately $150,000 for the software system and a software client cost of $25 for each STB.

Customer Premises Equipment (CPE)

The cost of the customer premises equipment is approximately $225 per STB.

Figure 8.3 shows a sample cost for adding IPTV service to a communication (such as a telephone system) that can provide service to 2,500 customers. The cost for head equipment includes $650,000 for 100 channel encoders and $250,000 for satellite reception equipment. The middleware system cost is $150,000 plus $25 for each middleware software (SW) client that is installed in each STB. Conditional access and DRM system cost is $150,000 plus $25 fore each CAS/DRM SW client that is installed in each STB. Each customer premises equipment STB costs $225. The overall capital cost per subscriber is $755.

Subscriber Information	Dedicated
Total # of Homes with Video Service	2,500
Dedicated Digital Head End Input:	
Price for 100 Channels of Digital Video Head End MPEG4 H.264 SD Encoders	$650,000
Price for IRDs, Cabinets and Cabling	$250,000
Middleware	
One Time Back Office System	$150,000
Base Client Fee (per set top box) - $25 each	$62,500
Conditional Access and DRM System	
One Time Back Office System	$150,000
Base Client Fee (per set top box) - $25 each	$62,500
Customer Premises Equipment	
Set Top Box - $225 each	$562,500
Total Capital Cost	**$1,887,500**
Capital Cost per Customer	**$755**

Figure 8.3, Sample Capital Costs for a Small to Medium Size IPTV System

Other Costs

There are many additional costs (some are hidden costs) associated with IPTV systems including financing, subscriber acquisition cost, post sales support, and billing systems.

Financing Cost

The cost to finance IPTV system equipment costs (debt load) can account for approximately 10-15 % of the service provider's cost. Many broadcasters and content service providers are highly leveraged with debt financing.

Subscriber Acquisition Cost (SAC)

Subscriber acquisition cost is the combined costs that are associated with marketing and adding of a customer to a system or service. Subscriber acquisition costs may include equipment subsidy, sales commissions, and associated marketing costs.

Post Sales Support

The sale of IPTV equipment involves a variety of costs and services after the sale of the product (post sales support), including technical support, warranty servicing, customer service, and training. A customer service department is required for handling distributor and customer questions. Because the average customer for an IPTV service is not technically trained in IPTV technology, the amount of non-technical questions can be significant.

Churn

Churn is the percentage of customers that discontinue service for any reason. Churn is usually expressed as a percentage of the existing customers that disconnect over a one-month period. Churn is often the result of natural migration (customers relocating) and the switching of service to other service providers. Because service providers commonly subsidize the cost of equipment that is provided to the customer and provide a service activation commission, churn can be a significant cost if the churn rate is high.

To help reduce the costs of churn, IPTV service providers may require the customer to sign a service agreement, which as a rule requires them to maintain service for a minimum period of time (commonly one year). These service agreements have a penalty fee in the event the customer disconnects service before the end of the one-year period.

Billing Systems

Each month, billing records must be totaled and printed for customer invoicing, invoices are mailed, and checks that are received are posted. The cost for billing services ranges approximately from $1 to $3 per month. Billing cost includes routing and summarizing billing information, printing the bill and the cost of mailing. To help offset the cost of billing, IPTV service providers may bundle advertising literature from other companies along with the invoice or provide online billing services. To expedite the collection, IPTV service providers may offer direct billing to bank accounts or charge cards.

References:

[1]. "Form 10Q, Time Warner Cable", Securities and Exchange Commission, www.SEC.gov, 7 Nov 2007.
[2] "US Retail E-Commerce Sales Maturing", eMarketer Research, www.eMarketer.com, 21 May 2007.
[3]. "Form 10Q, Time Warner Cable", Securities and Exchange Commission, www.SEC.gov, 7 Nov 2007.
[4]. "Form 10Q, Time Warner Cable", Securities and Exchange Commission, www.SEC.gov, 7 Nov 2007.
[5]. "Form 10Q, Time Warner Cable", Securities and Exchange Commission, www.SEC.gov, 7 Nov 2007.
[6]. "Form 10Q, Time Warner Cable", Securities and Exchange Commission, www.SEC.gov, 7 Nov 2007.
[7]. "IPTV Economics 101", Ed Coughlin, IPTV Magazine, March, 2006, www.IPTVMagazine.com.

Chapter 9

IPTV Challenges

Providing IPTV services may experience some challenges including content distribution rights, system capacity, content protection, television quality, reliability, regulation, privacy, channel changing time, patent licensing, HDTV, media player compatibility and security.

Content Distribution Rights

Content distribution rights are the authorized allowable methods that can be used by a content distributor (such as a broadcaster) to store, process and transfer content.

One of the key requirements for IPTV systems is to provide valuable content to its customers. Popular suppliers of content (such as television shows and movies) may limit their licensing to specific types of IPTV providers. An example of limited licensing is the licensing of local television channels only to cable television companies and not licensing to satellite broadcasting companies. Eventually government regulations in the United States forced local television program sources to fairly provide satellite broadcasters with local program content.

Because there are likely to be many IPTV service providers, content licensing for smaller IPTV providers is likely to occur through content aggregators. Content aggregation is the process of combining multiple content sources for distribution through other communication channels.

Content distribution rights typically have several key requirements including the term use of the material, how the material may be processed and where the content will be distributed. Licenses for content distribution that is based on the location of the user can be hard to control for IPTV system provider as the whereabouts of an Internet IPTV viewer may be anywhere in the world.

A content license is a contract that grants specific rights to use of content. Content owner license restrictions commonly include geographic region, type of use and amount of use. Content licenses may be setup in printed form and may not have detailed requirements transferred to content distributors. This can lead to the accidental omission of some licensing restrictions (such as authorized media uses). Content licensing automation systems are evolving to simplify the licensing and to help reduce these potential conflicts.

In addition to obtaining content licenses from the program source, program content may require additional licenses from many organizations for the distribution of certain media (such as music and video).

There are many new licensing risks for content licensing that established companies and newcomers need to prepare for. These risks include knowing what content rights are available to assign and obtaining content from companies that are not authorized to provide the content.

In general, the larger the number of media programs a company manages, the less they may know about the specific rights that may be available to them. Some IP rights may have already been assigned or the company may not know the specific rights they are authorized to have.

For companies that are new to licensing video content, they may not understand the risks and penalties associated with distributing unlicensed content. These companies may not perform due diligence (research) to confirm the rights are available to license from the company that is providing access to the content. There are hundreds of people and companies who offer content who may not actually have rights or may not have sufficient rights to distribute in this medium. The single biggest challenge to content licensing may be to ensure that the rights you think that you acquire are the rights that you actually own.

If content is distributed without a license, both the content distributor and/or access provider may be liable for copyright infringement. There is no statute of limitations on copyright infringement so legal disputes can appear 5 to 10 years after the copyright infringement has started with potentially devastating civil and even criminal penalties.

To protect against this risk, content distributors and access providers must perform due diligence (research) to confirm who actually owns the rights. Content aggregators due this on a regular basis and they may be able to insure and protect against potential licensing disputes that arise.

IPTV System Capacity

System capacity is the maximum information or service carrying ability of a communications system. The unit of capacity measurement for the facility or system depends on the type of services or information content that are provided by the system.

Actual deployments of IPTV systems seem to be dynamic and they will need to provide service to "hot spots" such as events and popular shows, both on the supply side (sports event sources) and on the distribution side (remote access requirements). As IPTV systems expand, they are likely to use VPNs from other companies to dynamically add capacity in hot spots.

When IPTV systems are integrated with data networks, the network capacity is shared between IPTV and data systems. The network capacity sharing issues include LAN network capacity and wide area network interconnection capacity.

Each television communication session for IPTV systems requires between 100 kbps (very low quality) to more than 4 Mbps of data transmission bandwidth (for standard television quality). The amount of bandwidth that each television viewing session requires varies based on the quality of service (QoS) desired and the type of data compression (video and audio compression) that is used. Because it is possible for the network operator to select the amount of data compression, this allows the network capacity to vary (called soft capacity) as it is possible for more users to view using the same amount of bandwidth by reducing the video and audio quality.

The sharing of bandwidth can be managed through the use of protocols (such as reservation protocols) or it can be left to the network to automatically decide how to allocate bandwidth. For high-speed data networks such as LAN systems, the amount of available bandwidth is usually tens or hundreds of times higher than the bandwidth required by each IPTV channel. As a result of this, the priority control processes designed into the standard data routers may be sufficient to provide for high QoS within the LAN system. For wide area networks that have more limited interconnection bandwidth capacity, priority can be assigned to data routers to for specific types of communication sessions (e.g. IPTV) to ensure bandwidth is available for voice communication.

Figure 9.1 shows how a data network shares bandwidth for both television and data communications. This diagram shows that a single router is providing data communications service to IPTVs and computer workstations. In this example, a computer workstation is transferring a large file and the IPTV is continuously streaming data at high-speed (2 – 4 Mbps). Because the LAN data network (Ethernet) has a maximum packet size of 1500 bytes of data and a standard high-speed data transmission rate of 100 Mbps, the router automatically divides the large file into smaller data blocks and access is shared between the IPTV and the computer workstation. When the data packets arrive at the relatively high-speed WAN connection, congestion can occur. If congestion were to occur, the router connected to the WAN connection would begin to delay the transmission of packets. In this example, the WAN router can give priority to the IPTV network packets and delays the file transfer packets.

Chapter 9

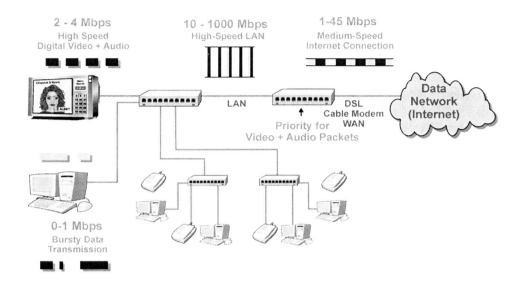

Figure 9.1, IPTV System Sharing Bandwidth

Content Protection

Content Protection is the end-to-end system preventing content from being pirated or tampered with in a communication network (such as in a television system). Content protection involves uniquely identifying content, assigning the usage rights, scrambling and encrypting the digital assets prior to play-out or storage (both in the network or end user devices) as well as the delivery of the accompanying rights to allow legal users to access the content.

Content protection can be divided into two functions; Conditional Access Systems (CAS) and Digital Rights Management (DRM). CAS controls access to the content and DRM controls how content can be distributed and used.

Content protection systems must give the IPTV service provider the ability to configure and control content rights while also ensuring that customer (and potentially employees of the IPTV service provider) cannot overcome these rights.

Because digital content can be reproduced multiple times without degrading the quality, content owners also desire to ensure that their content is protected. They also want to be assured that tracking systems correctly report usage to ensure they get paid. Content providers may want to see or perform revenue assurance tests.

Content providers may take an active role in reviewing and even defining requirements for CAS/DRM systems before agreeing to license their content to an IPTV service provider. If the IPTV service provider does not have CAS/DRM systems or has systems that are not acceptable to the content providers, getting programming content may be difficult or it may be limited to later release windows which have less market value.

Content Protection Options

There are two key types of content protection options; hardware (physical key) or software (soft key). Hardware solutions (such as smart card) have an advantage of being more difficult for hackers to break into and modify. However, once a hacker has succeeded, all of the hardware will need to be replaced. Software solutions can be reconfigured relatively quickly if they are compromised.

CAS and DRM systems require a software client to be integrated into the set top box. Because there are many manufacturers and models of STBs, integrating CAS/DRM software clients can be a complicated and difficult process.

Chapter 9

DRM systems are typically setup as client server systems where the system receives requests and provides services to clients. A DRM system server may include the content server (the content source), content descriptions (metadata), DRM packager (media formatter), license server (rights management), and a DRM controller (DRM message coordinator). A DRM client typically includes a DRM controller, security interface (key manager) and a media decoder.

A license server is a computer system that maintains a list of license holders and their associated permissions to access licensed content. The main function of a license server is to confirm or provide the necessary codes or information elements to users or systems with the ability to provide access to licensed content. The license server may download a key or other information to client devices that enables a license holder to access the information they have requested.

License servers use licensing rules to determine the users or devices that have authorization to access data or media. Licensing rules are the processes and/or restrictions that are to be followed as part of a licensing agreement. Licensing rules may be entered into a digital rights management (DRM) system to allow for the automatic provisioning (enabling) of services and transfers of content.

A key server is a computer that can create, manage, and assign key values for an encryption system. A key is a word, algorithm, or program that used to encrypt and decrypt a message that is created in a way that does not allow a person or system to discover the process used to create the keys.

DRM systems may have the capability to transfer and update keys (key renewability). Key renewability is the ability of an encryption system to issue new keys that can be used in the encoding or decoding of information.

A DRM controller is the coordinator of software and/or hardware that allows users to access content through a digital rights management system. DRM controllers receive requests to access digital content, obtain the necessary information elements (e.g. user ID and key codes), performs authentication (if requested) and retrieves the necessary encryption keys that allows for the decoding of digital media (if the media is encoded).

A digital rights management client is an assembly, hardware device or software program that is configured to request DRM services from a network. An example of a DRM client is a software program (module) that is installed (loaded) into a converter box (e.g. set top box) that can request and validate information between the system and the device in which the software is installed.

Studio Endorsements

A studio endorsement is the allowance of use of a program or service as long as the use meets with specific distribution requirements. Unfortunately, studios do not typically endorse (authorize) the use of specific content protection as CAS/DRM programs may not meet all of their security requirements.

There are two types of security systems, the ones that are secure and the ones that have not been compromised yet. Security systems are constantly evolving and implementing CAS/DRM systems is a continuous process. Content providers are likely to be interested in the existing CAS/DRM system and the upgrade and evolution plans for content protection systems.

Because content is the key value product for studios, they have developed knowledge of content protection issues and systems. As a result, studios and content providers can be very demanding on understanding the CAS/DRM technology that is to be used to protect their content.

Of particular concern to content providers is what happens when the technology is compromised. If the system is compromised, there should be a recovery plan.

Television Quality

Television networks provide a fairly high level of quality of service (QoS) to television viewers and to be successful, IPTV service should have similar quality, security and reliability as existing television systems.

Audio Quality

Audio quality is the ability of a speaker or audio transmission system to recreate the key characteristics of an original audio signal. Audio quality can be affected by a variety of factors. These factors include the type of audio coder (audio compression), transmission system and bandwidth limits.

Generally, the more you compress the audio, the lower the audio quality. Recently, innovations in audio compression technology provide similar quality audio signals using a much lower data communication (connection) speed.

IPTV service can provide audio quality that is the same or better than standard television quality audio. IPTV systems can offer advanced audio options including stereo and surround sound.

Audio distortion is the undesired changing of an audio signal and it can come from a variety of sources in IPTV service. However, some of the key factors in audio distortion are packet loss and packet corruption.

Packet loss is the inability of the communication network to deliver a packet to its destination within a prescribed period of time. The effect of packet loss on audio distortion is to temporarily mute or distort the audio signal. Packet loss can result from a variety of events such as network congestion or equipment failures.

Because IPTV communication systems can delay (buffer) the delivery of packets of data, it is usually possible and practical to resend packets of data that contain audio information. As a result, packet losses are infrequent. When packets are lost, this can result in the temporary muting of audio.

Packet corruption is the changing of some of the packet data during its transmission. Packet corruption can come from a variety of sources such as poor communication line quality or momentary line loss from lightning spikes. Because IPTV service typically uses speech compression, the packet data represents a sound that will be recreated rather than a specific portion of the actual audio signal. As a result, if corrupted data is used, this can create a very different audio sound then expected. This distorted sound is commonly called "Warble."

Figure 9.2 shows some of the causes and effects of audio distortion in IPTV systems. This example shows that audio signals are digitized, compressed and error protection coded prior to transmission. During the transmission process, some packets are lost or corrupted. The loss of packets can result in

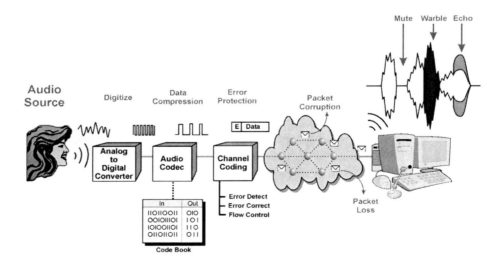

Figure 9.2, IPTV Audio Distortion

the temporary muting of the audio signal. Because the data compression process represents sounds by different codes in a codebook, packet corruption results in the creation of a different altered sound than the sound that was previously transmitted. When there is significant data corruption, this can create unusual sounds (a "Warble" sound).

Video Quality

Video quality is the ability of a display or video transfer system to recreate the key characteristics of an original video signal. Traditional video quality impairment measurements include blurriness and edge noise. Digital video and transmission system impairments include tiling, error blocks, smearing, jerkiness, edge busyness and object retention.

Video quality can be affected by a variety of factors that interact with each other. Some of these factors include the choice of video coder (video compression), transmission type and bandwidth limitations. The types of distortion on analog video systems include blurriness and edge noise. Digital video and transmission system impairments include tiling, error blocks, smearing, jerkiness, edge busyness and object retention.

Tiling is the changing of a digital video image into square tiles that are located in positions other than their original positions on the screen. Error blocks are groups of image bits (a block of pixels) that do not represent error signals rather than the original image bits that were supposed to be in that image block. Jerkiness is holding or skipping of video image frames or fields. Object retention is the keeping of a portion of a frame or field on a display when the image has changed.

Figure 9.3 shows some of the causes and effects of video distortion that may occur in IPTV systems. This example shows that video digitization and compression converts video into packets that can be sent through data networks (such as the Internet). Packet loss and packet corruption results in distorted video signals. This example shows that some types of digital video distortion include tiling, error blocks and retained images.

IPTV Business Opportunities

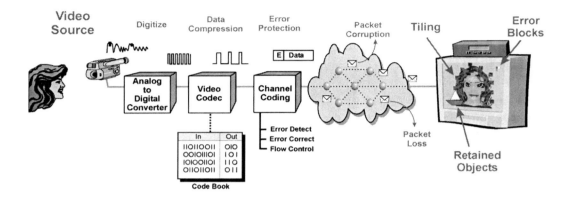

Figure 9.3, IPTV Video Distortion

Reliability

Reliability is the ability of a network or equipment to perform within its normal operating parameters to provide a specific quality level of service. Reliability can be measured as a minimum performance rating over a specified interval of time. These parameters include bit error rate, minimum data transfer capacity or mean time between equipment failures (MTBF).

Reliability factors for IPTV service includes IPTV access device reliability, data network connection reliability, data network reliability, call server reliability and feature operation reliability.

Access Device Reliability

Access device reliability is the ability of device or system equipment to allow a user to gain access to a network within a specific quality level of service. For IPTV service, the access device must be able to request and view IPTV channels. To be effective, IPTV access devices must be able to continuously process video and audio signals during the viewing of a channel. Access device operation may be dedicated (such as an IPTV) or they may be shared (such as viewing on a multimedia computer).

Access devices are often connected to a modem or local data network equipment. The reliability of these local data communication devices also affects the reliability of IPTV service. Some of these devices may change their configuration during connection and disconnection. If the data communication device does not appear to be working, it is best to turn its power off and restart the equipment.

Figure 9.4 shows that the selection of access device can affect the operation and quality of IPTV service. In this example, a standard television that has an IPTV set top box (analog television adapter) and a laptop computer are viewing a television channel via a media server through the Internet. The analog television adapter is designed to for viewing IPTV service and it always has the resources (processing power) to do this. Unfortunately, the

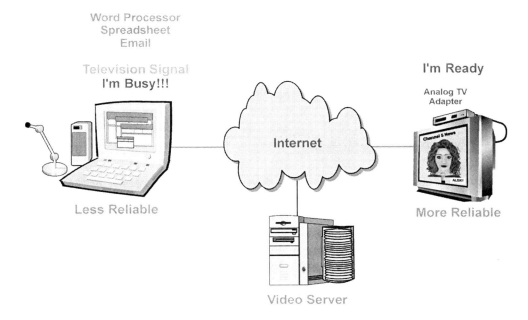

Figure 9.4, IPTV Access Device Reliability

laptop computer is a multipurpose device that is currently running several applications (word processor, spreadsheet, and email). When the laptop computer receives this television channel, the other processes may cause the audio and video to become somewhat distorted or the resulting delays may drop the media connection.

Data Network Reliability

Data network reliability is the ability of the communication network to consistently provide data transmission between points that are connected to the data network. Data networks such as the Internet were designed to successfully operate even if large portions of the network were destroyed. To accomplish this, the Internet was designed as a dumb network that uses smart switches. Each switch in the Internet (called a router) has the ability to dynamically change the path it uses to sending data through based on information it regularly receives from other routers. If a router can no longer send data to a neighboring router, it will automatically start to send data to a router it can communicate with. As a result, the Internet is very reliable as it can repair itself in the event equipment failures.

Figure 9.5 shows that the Internet is a web of paths that interconnect endpoints and that if this web is broken, it is possible for information to take another path to reach its destination. This rerouting of information is automatic.

Data Connection Reliability

Data connection reliability involves the connection from an access device (such as a computer or IP set top box) to a data network (such as the Internet). Your data connection may be divided into two parts; access provider and data network provider (such as an ISP). The access provider manages the connection between your equipment and the data network provider converts your data into a format that it can transmit through the data network.

Figure 9.6 shows the key parts of an Internet service provider (ISP) and how they can affect your communications reliability. This diagram shows that an Internet connection can be divided into an ISP portion and an access

Chapter 9

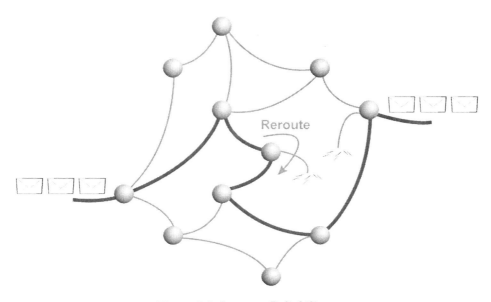

Figure 9.5, Internet Reliability

provider portion. This example shows an IPTV that is connected to a cable modem. The cable modem is connected to the head-end of the cable television system where a gateway adapts the data from the cable network into a format that can be used by the ISP. The ISP has a router that connects the gateway into a format that is sent to the Internet. This diagram shows that this ISP only has on connection to the Internet and if it experiences difficulty, the Internet connection can be lost.

IPTV Business Opportunities

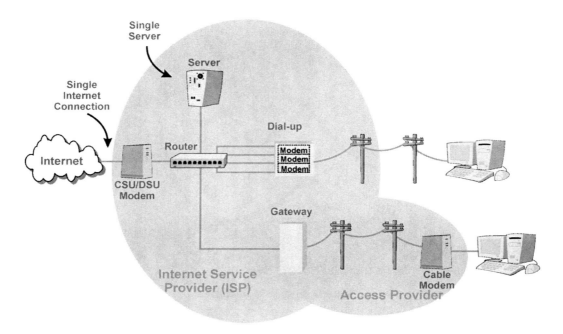

Figure 9.6, Data Connection Reliability

IPTV Server Reliability

IPTV server reliability is the ability of a television service provider (television channel processing system) to setup and control IPTV channels along with selecting and managing video switches and gateways. To ensure reliability, IPTV service providers may have redundant (duplicate) server equipment, updated lists of IPTV gateways and use equipment that confirms to specific and compatible revisions of communication protocols.

Figure 9.7 shows the key parts of an IVTSP that is used to provide IPTV service and how the configuration can affect reliability. In this example, the ITVSP television server has two media centers that are connected to the Internet at different locations. IPTVs communicate with the ITVSP servers to setup and receive television channels. Each ITVSP sever has a media

Chapter 9

provider list that comes from a company that maintains lists of the gateways at media sources (e.g. movie distributors). In the event of a failure of one of the servers, the other server will operate to setup and connect IPTV channels.

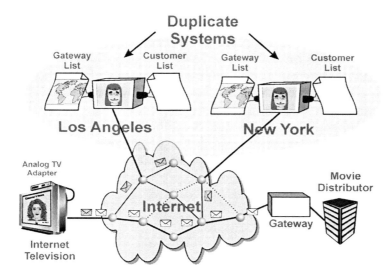

Figure 9.7, ITVSP Reliability

Feature Operation Reliability

Feature operation reliability is the ability of the system to recognize and process feature requests. There are many features available in television networks (such as pay per view, pause, rewind and mute) and these features have been designed and tested to interoperate with each other. These features are usually managed by a single system. When these features are offered via IPTV service providers, there may be interaction with these features with features offered by different service providers. This can cause challenges with the operation of specific features. For example, if an ITVSP provides access to movies on demand and the media server uses a proprietary protocol for control messages, the player control commands or menu selection may not work correctly.

175

Regulation

Regulatory issues for IPTV include franchise restrictions, fair access to content, control of content distribution and taxes. Cable and television broadcasters have been commonly given franchise rights to exclusively distribute television services in geographic areas and these franchise agreements may be interpreted to limit the ability for IPTV providers to offer services customers. IPTV system operators need to have fair access to local and network programming and it may not in the business interests of other broadcasters to provide access to their content. Certain types of content such as programs that contain pornography, gambling or violence may have distribution restrictions in geographic regions and broadband TV providers may have difficulty controlling where people watch the programs. New taxes are likely imposed on IPTV operators and how these taxes are assessed and collected may be hard to perform.

Privacy Requirements

Privacy requirements are the regulatory and business rules for the restriction or presentation formats of information that is provided to people or companies other than those that own or have rights to obtain the information. Privacy laws are regulatory requirements that restrict the use and transfer of information that is considered as private to person, company or owner of the information. Violations of privacy can cause embarrassment or financial losses.

A key potential challenge for IPTV privacy is the linking of relatively unimportant or unusable information to create more useful information that can violate the privacy of individuals. For example, if one order includes only the last 4 digits of an 8 digit account number, this may not be very valuable. However, if another invoice were linked that included the first 4 digits of the account number, the entire account number has been obtained.

Some of the privacy issues include protecting the numbers, connections, or media requested or accessed by individuals. IPTV service providers may need to protect the identity and location of users who do not want to be listed in directories.

Privacy can mean different things to different people. Companies can have a privacy policy or privacy statements to inform customers of the potential uses of customer information. Privacy policies are the self proclaimed rules a receiver of information claims to follow when a customer or visitor sends or provides information. Privacy policy rules typically state how the information may be used and who the information may be distributed to.

HDTV over IPTV

Television service providers need to consider several key options when offering HDTV and other high bandwidth television services (such as 3DTV) over their IPTV systems. If providing many choices of HDTV programming is an objective of the service provider, IPTV is one of the most reasonable ways to provide access to HDTV programming. A traditional television broadcast system has a limited amount of capacity and if many of the programming channels were converted to HDTV, the broadcaster would need to reduce the number of available channels. Once an IPTV system is setup to offer HDTV, adding additional HDTV programs is not an issue.

Patent Licensing

Patent licensing is the process of obtaining the rights to use technology that is described in one or more patents. A patent license is an authorization to produce, use, or sell products that may use technologies or processes defined by the claims in a patent.

As the IPTV industry continues to grow, attracting increasing numbers of subscribers and investors, it may also attract the attention of the companies that have patents relating to components of IPTV (such as streaming media).

Media Player Compatibility

Media player compatibility is the capability of a media processing application (such as Windows Media Player) to receive, convert and control media such as video, audio or images into a form that can be experienced by humans. Media players may contain support for service different media formats, compression (codec) formats as well as being able to communicate using multiple network streaming protocols.

In addition to the specific digital audio, digital video and control protocol capabilities, there may be multiple versions of each capability. Later versions of media players may attempt to provide or use features that are not available or fully developed in the other media player. This may cause unexpected results such as audio muting, display format changes or other changes in desired operation.

Because each media player has multiple capabilities, media players may negotiate their preferred media options. The process of requesting and agreeing on the preferred characteristics for a communication session in parameter negotiation.

Figure 9.8 shows how a media player is composed of video and audio coders along with control protocols. The media server and media player may negotiate on which coders and protocols to use during their communication session. This example shows that the media server desires to use MPEG-4. Because the media player does not have MPEG-4 capability, it responds that it desires to use MPEG-2 and the media server accepts the request. The media server desires to use the AAC audio coder and the media player accepts.

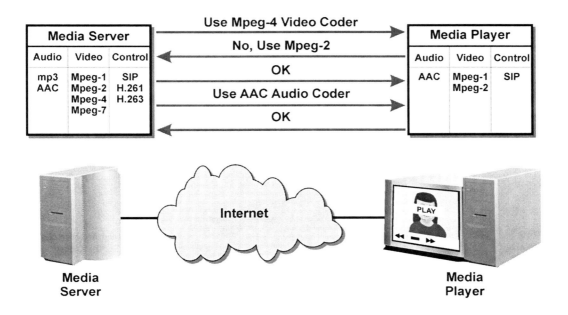

Figure 9.8, Media Player Compatibility

Channel Change Time

Channel change time is the duration between when a channel change request is initiated (e.g. pressing a channel button) and when the display of the selected channel video begins. Channel changing time is delayed primarily due to the buffering of data packets. Packet buffering is used to overcome the variable delays that naturally occur in packet data networks.

Some of the ways to decrease the channel changing time include dramatically increasing the bandwidth, increasing the compression during the beginning of the play mode, or providing the video image in a lower quality format during the initial channel changing time.

IPTV Business Opportunities

Figure 9.9 shows how media servers and transmission buffering can cause increases in IPTV channel changing time. This example shows that the time period between requesting a channel change to when the new channel starts can depend on the media server processing time and the buffering time in the media player. This example shows that a media server must receive the request and obtain the media source (such as a DVD) and begin streaming the media to the viewer's media player. When the media player begins to receive the new channel, it will buffer information to help ensure gaps or distortion is not likely to occur in the playing of the new channel.

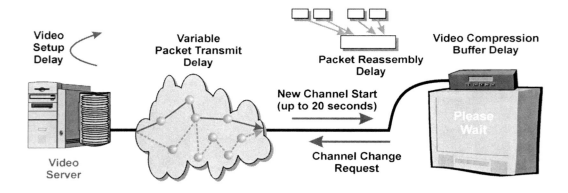

Figure 9.9, IPTV Channel Changing Time

IPTV Security

A growing awareness of security seems to be present. Because IPTV uses standard IP protocols, IPTV systems face some of the same security threats as web systems. This is especially true as IPTV systems shift from using private networks to public infrastructure. A key security threat to IPTV systems is the use of protocol spidering. Protocol spidering is the use of different combinations of commands and parameters (protocol commands) in an attempt to gain unauthorized access into networks or the ability to control or modify the operations of devices or services.

Chapter 10

IPTV System Integration

If you are deploying or have already deployed IPTV systems, there are many challenges that you have or will encounter such as defining requirements, selecting and purchasing equipment, integrating systems and testing. Companies use system integrators to help design, setup and maintain IPTV systems. By choosing the right system integrator, many of these challenges can be solved. In addition, the use of system integrators can save significant capital and operational costs and revenue by identifying new services and applications.

System integration can be a complex process. While many systems work on a small scale deployment, as systems grow, key congestion points appear, especially during high-activity periods (e.g. channel zapping during popular sports games). Testing while simulating fully loaded conditions can be vital during IPTV system deployment.

Further complicating the situation is the growing number of available IPTV devices. Many IPTV products have evolved to their 2nd and 3rd generation versions and within each version there can be multiple software revisions. There are also many new manufacturers of set top boxes who desire to test for interoperability.

A key benefit of the growth of the industry is that prices are dropping rapidly as volumes increase. Some STB manufacturers are producing over 500,000 units per year. Companies who specialize in products for specific industries such as for Hotels and Universities are also deploying IPTV systems. These vertical channels of distribution have the potential for large orders which can have a relatively simple sales process.

Systems Integrators

There are several situations that may occur to motivate IPTV systems operators to consider using system integrators. Companies that are deploying new IPTV systems may decide to use system integrators because they do not possess the necessary skills or experience to design and deploy a system. Some companies that have started to deploy IPTV systems begin to run into interoperability problems and working with multiple vendors proves to be complicated. In some cases, no resolution to multi-vendor problems can be achieved.

System integrators can provide technical and business skill sets that the operator may not yet have or that they only need to use during the design and setup phases. System integrators have already discovered and overcome the many undocumented problems that occur when setting up and interconnecting equipment supplied by multiple vendors. This experience can reduce the number of problems and mistakes that can occur, which reduces implementation delays and lowers development costs.

System integrators have also been exposed to many options available from multiple vendors. These options can translate into lower cost designs (e.g. less redundancy) and the potential for some new revenue producing services (such as ad insertion) that may not have been factored into business plans. The system operator may also be able to provide an unbiased perspective of these options, reducing the confusion when multiple vendors may overstate the benefits of solutions and avoid the discussion of potential problem areas.

IPTV system operators may be able to leverage volume purchasing discounts made available to systems integrators. The system integrators have typically pre-validated equipment selection to avoid the installation and setup of troubled equipment or incompatible system configurations.

One of the most important benefits may be the single point of contact that system integrators provide to operators. This can avoid or reduce multi-vendor challenges where each vendor claims their system is not the problem finger pointing the problem to other vendors.

Figure 10.1 shows some of the reasons why companies decide to use IPTV systems integrators.

Reason	Notes
Obtain Skill Sets	Technical and Business. May be Only Temporarily Needed
Make Less Mistakes	Reduced Delays and Higher Reliability
Additional Options	Lower Cost Implementations and Additional Types of Revenue Services
Purchasing Discounts	Equipment Discounts due to Vendor Relationships
Pre-Validate Equipment and Services	Avoid Problem Equipments
Single Point of Contact	Single Point of Contact. No Finger Pointing

Figure 10.1, Reasons to Use System Integrators

Qualifying an IPTV Systems Integrator

IPTV systems integrator qualifications range from basic company credibility checks to experience with specific types of services and systems.

System integrators should be able to work on all types of systems and equipment including access networks, video servers and software applications. The systems integrator should be able to describe the number and types of systems they have integrated along with highlighting some of the many integration issues that will occur when mixing products from different vendors. They should also have experience in working with underlying systems such as optical distribution and DSL access systems.

The systems integrator should have support systems including call centers that is staffed with employees (not outsourced from other companies) that have the necessary skill sets to support the many aspects of IPTV system operation ranging from testing to billing systems.

System integrators usually have good vendor relationships and they purchase quantities of equipment from these suppliers entitling them to volume discounts. These cost reductions can be passed on to the IPTV operator which may be purchasing smaller quantities of products. Good vendor relationships can also lead to better communication with suppliers to solve complicated problems (e.g. knowing the person in charge of specific products or subassemblies).

Setting up IPTV systems involves a mix of hardware and software engineering capabilities. The systems integrator should have training and qualified staff that has practical design experience along with industry certifications. There should be established teams of installers which hold the necessary licenses and certifications.

Systems integrators should have test facilities and testing capabilities that can support the evaluation, performance and acceptance tests that are necessary for the setup of the IPTV systems. The systems integrator should have training capabilities such as on site, off site and online courses. The training courses should have good supporting materials and laboratory exercises when appropriate.

Figure 10.2 shows a quick checklist of attributes that may be used to qualify an IPTV systems integrator.

Attribute	Notes
Experience	Installed and operated systems with similar size, services and applications
Support	Can be a source of solution information
Vendor Relationships	Works with multiple vendors. May be able to obtain equipment and software discounts
Engineering	Has certified hardware and software designers
Installation	Has team of installers and supporting installation equipment
Testing	Has test laboratory and test procedures
Training	Has both on-site and off-site training capabilities

Figure 10.2, IPTV System Integrator Check List

Appendix 1 - Acronyms

AAC-Advanced Audio Codec
ADSL-Asymmetric Digital Subscriber Line
AoD-Advertising on Demand
ARPU-Average Revenue Per User
BBTV-Broadband TV
BPL-Broadband Over Powerline
CA-Conditional Access
CapEx-Capital Expenditure
CAS-Conditional Access System
CATV-Cable Television
CC-Closed Caption
CM-Cable Modem
CMTS-Cable Modem Termination System
CPE-Customer Premises Equipment
DNS-Domain Name Server
DRM-Digital Rights Management
DSL-Digital Subscriber Line
DSLAM-Digital Subscriber Line Access Multiplexer
DVB-H-Digital Video Broadcasting Handheld
DVQ-Digital Video Quality
DVR-Digital Video Recorder
E-commerce or ECommerce-Electronic Commerce
Enterprise TV-Enterprise Television
EPG-Electronic Programming Guide
EVDO-Evolution Version Data Only
GPRS-General Packet Radio Service
GW-Gateway
HD-High Definition
HDTV-High Definition Television
HomePNA-Home Phoneline Networking Alliance
IAB-Interactive Advertising Bureau
IGMP-Internet Group Management Protocol
IP Address-Internet Protocol Address
IP STB-Internet Protocol Set Top Box
IPCATV-Internet Protocol Cable Television
IPTV-Internet Protocol Television
ISP-Internet Service Provider
ITU-International Telecommunication Union
iTV-Internet TV
ITVSP-Internet Television Service Provider
Linear TV-Linear Television
LMDS-Local Multichannel Distribution Service
LTV-Lifetime Value
MAC-Medium Access Control
M-Commerce-Mobile Commerce
MDU-Multiple Dwelling Unit
MG-Media Gateway
MG-Minimum Guarantee

MMDS-Multichannel Multipoint Distribution Service
MP3-Motion Picture Experts Group Layer 3
MP4-MPEG-4
MPEG-Motion Picture Experts Group
MS-Media Server
MSB-Multiple Service Bundle
MSO-Multiple System Operator
MTBF-Mean Time Between Failures
MTU-Multi Tenant Unit
MWS-Multimedia Wireless Systems
NTSC-National Television System Committee
Off Net-Off Network
On Net-On Network
OpEx-Operational Expenses
PAL-Phase Alternating Line
PDN-Premises Distribution Network
PLC-Power Line Carrier
PMC-Personal Media Channel
PPV-Pay Per View
PVOD-Push Video on Demand
QoS-Quality Of Service
Reciprocity-Reciprocal Compensation
SAC-Subscriber Acquisition Cost
SD-Standard Definition
SDSL-Symmetrical Digital Subscriber Line
SDTV-Standard Definition Television
SG-Signaling Gateway
SIP-Session Initiation Protocol
STB-Set Top Box
Sublicensing-Sub-Licensing
SVS-Switched Video Service
T-commerce-Television Commerce
TelcoTV-Telephone Company Television
TV Portal-Television Portal
TV Set-Television Set
UGC-User Generated Content
URM-User Rights Management
VDSL-Very High Bit Rate Digital Subscriber Line
VOD-Video On Demand
VS-Video Server
WBB-Wireless Broadband
WCDMA-Wideband Code Division Multiple Access
WikiTV-Wiki Television
WiMax-Worldwide Interoperability for Microwave Access
WLAN-Wireless Local Area Network
XSLT-XML Stylesheet Language Transformation

Appendix 2 - Research Companies

A.C. Nielson

www.acnielson.com

AccuStream iMedia Research

www.accustreamresearch.com/
AccuStream iMedia Research
1401 Madrone Drive
Salinas, CA 93905
Phone: +1-831-757-2556
Fax: +1-831-757-2553

eMarketer

www.emarketer.com/
eMarketer
75 Broad Street
32nd Floor
New York, NY 10004
Phone: +1-212-763-6010
800-405-0844 (Toll-free)
Fax: +1-212-763-6020

Forrester Research

www.forrester.com/
Forrester Research, Inc.
400 Technology Square
Cambridge, MA 02139 USA
Phone: +1-617/613-6000
Fax: +1-617/613-5200

Infonetics

www.infonetics.com/
900 E. Hamilton Ave
Suite 230
Campbell, CA 95008
Phone: +1-408.583.0011
Fax: +1-408.583.0031

Instat

www.instat.com
2055 Gateway Place, Suite 150,
San Jose CA 95110
Phone: +1-408.345.4495

International Telecommunications Union

www.itu.int
International Telecommunication Union (ITU)
Place des Nations
1211 Geneva 20
Switzerland
Phone: +41 22 730 5111
Fax: +41 22 733 7256

Maravedis

www.maravedis-bwa.com
410 Rue des Recollets Suite 301
Montreal, Quebec, H2Y 1W2
Canada
Phone: +1-305 992-3196
Fax: +1-514 313-5465
info@maravedis-bwa.com

MRG

www.mrg.com/
Management Research Group, Inc.
14 York Street, Suite 301
Portland, Maine 04101 USA
Phone: +1-207 775-2173
Fax: +1-207 775-6796

Point Topic

www.point-topic.com
Point Topic Ltd,
175 Gray's Inn Road,
London WC1X 8UE,
United Kingdom.
Phone: +44 (0) 20 7812 0506
Fax: +44 (0) 20 7812 0650

Interactive Advertising Bureau
www.IAB.net
116 East 27th Street, 7th Floor
New York, New York 10016
Phone: +1-212 380 4700

Vision Gain

www.visiongain.com
Visiongain Ltd.
4th Floor,
BSG House,
226-236 City Road,
London
EC1V 2QY
United Kingdom
Phone: +44 (0) 20 7336 6100
Fax: +44 (0) 20 7549 9930

Index

Access Device Reliability, 170-171
Ad Telescoping, 41-42
Addressable Advertising, 13, 34, 40, 91-92
Advanced Audio Codec (AAC), 82, 178
Advertising, 13, 20, 27-28, 30-31, 34-43, 53, 55, 64, 66-67, 71, 91-92, 143, 145-148, 157
Advertising Rights, 66, 147
Applied Metadata, 73
Asymmetric Digital Subscriber Line (ADSL), 110, 114
Audio Compression, 77, 162, 167
Audio Quality, 80, 162, 167
Average Revenue Per User (ARPU), 6, 10, 23, 43, 145
Bandwidth Capacity, 162
Billing System, 36, 47
Brandable, 69
Broadband Over Powerline (BPL), 52
Broadband TV (BBTV), 63, 135, 137, 176
Bundled, 145
Cable Modem (CM), 7-9, 29, 50, 88, 101, 116-117, 131, 173
Cable Modem Termination System (CMTS), 50, 117
Capital Costs, 153, 155
Capital Expenditure (CapEx), 138

Channel Change Time, 179
Channel Changing Time, 159, 179-180
Churn, 156-157
Community Content Programming, 61-62
Conditional Access (CA), 99, 154-155, 163
Conditional Access System (CAS), 99, 163
Content Acquisition, 61, 69, 103, 136
Content Aggregator, 3
Content Consumer, 3
Content Distribution Rights, 159-160
Content Distributor, 3, 60, 159, 161
Content Licensing, 64, 66, 135, 160-161
Content Lifecycle, 71
Content Partner, 69
Content Producer, 3
Content Provider, 62, 71
Core Metadata, 73
Customer Premises Equipment (CPE), 155
Data Compression, 76, 162, 169
Data Connection Reliability, 172, 174
Data Network Reliability, 170, 172

Digital Rights Management (DRM), 99-102, 137, 154-155, 163
Digital Subscriber Line (DSL), 7-8, 13, 50, 88, 108, 113-115, 185
Digital Subscriber Line Access Multiplexer (DSLAM), 114-115
Digital Video Broadcasting Handheld (DVB-H), 20, 22, 52
Digital Video Recorder (DVR), 5, 33
Digitization, 75-76, 169
Distribution Rights, 159-160
Download and Play, 93-95
Electronic Commerce (E-commerce or ECommerce), 23
Electronic Programming Guide (EPG), 90
Encryption, 99, 101, 141, 165
Enterprise Television (Enterprise TV), 67
Evolution Version Data Only (EVDO), 20, 121
Fiber Access Line, 11
Flat Tail Content, 72
Gateway (GW), 33, 56, 71, 82, 84, 86, 132, 134, 173
General Packet Radio Service (GPRS), 20
Headend, 2, 103, 120, 125-126, 136, 154, 173
High Definition (HD), 15, 76, 94-95, 135, 140
High Definition Television (HDTV), 1, 11, 51, 159, 177
Home Phoneline Networking Alliance (HomePNA), 110
Interactive Advertising Bureau (IAB), 27, 30
Interactive Applications, 34, 139

International Programming, 63-64, 66
International Telecommunication Union (ITU), 14, 30, 121
Internet Protocol Address (IP Address), 89, 108, 133
Internet Protocol Set Top Box (IP STB), 15, 108
Internet Service Provider (ISP), 100, 172-173
Internet Television Service Provider (ITVSP), 83, 98, 100-101, 132-134, 174-175
Internet TV (iTV), 49, 54-55, 66, 131
Interstitial Ad, 35
IPTV Advertising, 35-36, 38, 40, 53
IPTV Network, 98, 125, 135-136, 162
Jitter, 81
Key Players, 49
Linear Television (Linear TV), 31, 63, 71, 150
Local Multichannel Distribution Service (LMDS), 114, 118-120
Local Programming, 60-61, 63
Long Tail Content, 72
Managed IPTV, 54, 131, 135, 137
Mean Time Between Failures (MTBF), 170
Media Compression, 76-77
Media Consumption, 4, 21-22
Media Consumption Habits, 4, 21
Media Format, 94
Media Gateway (MG), 56, 82, 134
Media Player Compatibility, 159, 178-179

Index

Media Server (MS), 81-82, 84, 86, 89, 96, 134, 171, 175, 178, 180
Medium Access Control (MAC), 116, 123
Metadata Management, 73
Middleware, 145, 154-155
Mixed Media, 35, 43-44, 68
Mobile Commerce (M-Commerce), 23
Motion Picture Experts Group (MPEG), 82, 105
MPEG-2, 77, 178
MPEG-4 (MP4), 15, 77, 178
Multicast, 52, 86-88
Multichannel Multipoint Distribution Service (MMDS), 114, 118-120
Multimedia Wireless Systems (MWS), 120
National Television System Committee (NTSC), 82, 108, 116
Network Capacity, 97, 153, 161-162
Network Operator, 35, 60, 69, 100, 135-137, 162
Network Programming, 60, 138, 146, 149, 176
Off Network (Off Net), 134
On Network (On Net), 134
Original Programming, 50, 60
Overlay Ad, 39
Packet Buffering, 80-81, 179
Packet Corruption, 167-169
Packet Losses, 79-81, 168
Packet Routing, 1, 78
Paid Placement, 36
Patent License, 177
Patent Licensing, 159, 177

Pay Per View (PPV), 18-21, 31, 146, 175
Personal Media Channel (PMC), 68-69
Personalized Advertising, 41
Phase Alternating Line (PAL), 82, 108, 116
Power Line Carrier (PLC), 123
Private IPTV, 131, 139-143
Program Guide, 89
Progressive Downloading, 96-97
Push Video on Demand (PVOD), 97
Quality Of Service (QoS), 94, 97, 111, 123, 132, 136, 162, 167
Rate Plan, 16, 19
Reciprocal Compensation (Reciprocity), 152
Recommendation Engine, 91
Reliability, 131, 140, 159, 167, 170-175
Remote Diagnostics, 140
Remote Monitoring, 140
Roaming, 33, 51, 55, 121
Rules Based Advertising, 39
Session Initiation Protocol (SIP), 100, 105
Set Top Box (STB), 3, 15, 19, 32, 84, 95, 97, 108, 115, 134, 138-140, 142, 154-155, 165-166, 171-172, 183
Shared Content, 21, 62
Shared Headend, 154
Short Tail Content, 72
Sponsored Content, 61
Standard Definition (SD), 59, 76, 94-95, 135
Standard Definition Television (SDTV), 59

Streaming, 22, 65, 71, 81, 83, 86, 95-96, 162, 177-178, 180
Subscriber Acquisition Cost (SAC), 156
Switched Video Service (SVS), 84
Switching, 40, 49, 55, 83-84, 89, 115, 135, 154, 156
Symmetrical Digital Subscriber Line (SDSL), 114
System Integration, 183
Target Market, 34
Telephone Company Television (TelcoTV), 131
Telephone Lines, 13-14, 109-110, 135
Television Advertising, 13, 147
Television Commerce (T-commerce), 23, 31, 42-47, 145, 147-148
Television Set (TV Set), 97
Time Offset, 64-65
Transactional Metadata, 73
Ultra Broadband, 112
Unicast, 86-87
Untrusted Device, 100
User Generated Content (UGC), 28-29
User Rights Management (URM), 99-100
Very High Bit Rate Digital Subscriber Line (VDSL), 114
Video Catalog, 43
Video Compression, 34, 77, 169
Video On Demand (VOD), 16, 92-93, 97, 108, 120, 140, 164
Video Quality, 131, 169
Video Server (VS), 28
Wholesale On Demand, 69-70

Wideband Code Division Multiple Access (WCDMA), 121
Wiki Television (WikiTV), 62
Wireless Broadband (WBB), 7-11, 49, 53-54, 113, 118-120, 131
Wireless Local Area Network (WLAN), 111, 114, 122-123
Worldwide Interoperability for Microwave Access (WiMax), 8-10, 51, 54, 123, 129, 136

Althos Publishing Book List

Product ID	Title	# Pages	ISBN	Price	Copyright
Billing					
BK7781338	Billing Dictionary	644	1932813381	$39.99	2006
BK7781339	Creating RFPs for Billing Systems	94	193281339X	$19.99	2007
BK7781373	Introduction to IPTV Billing	60	193281373X	$14.99	2006
BK7781384	Introduction to Telecom Billing, 2nd Edition	68	1932813845	$19.99	2007
BK7781343	Introduction to Utility Billing	92	1932813438	$19.99	2007
BK7769438	Introduction to Wireless Billing	44	097469438X	$14.99	2004
IP Telephony					
BK7781361	Tehrani's IP Telephony Dictionary, 2nd Edition	628	1932813616	$39.99	2005
BK7781311	Creating RFPs for IP Telephony Communication Systems	86	193281311X	$19.99	2004
BK7780530	Internet Telephone Basics	224	0972805303	$29.99	2003
BK7727877	Introduction to IP Telephony, 2nd Edition	112	0974278777	$19.99	2006
BK7780538	Introduction to SIP IP Telephony Systems	144	0972805389	$14.99	2003
BK7769430	Introduction to SS7 and IP	56	0974694304	$12.99	2004
BK7781309	IP Telephony Basics	324	1932813098	$34.99	2004
BK7780532	Voice over Data Networks for Managers	348	097280532X	$49.99	2003
IP Television					
BK7781334	IPTV Dictionary	652	1932813349	$39.99	2006
BK7781362	Creating RFPs for IP Television Systems	86	1932813624	$19.99	2007
BK7781355	Introduction to Data Multicasting	68	1932813551	$19.99	2006
BK7781340	Introduction to Digital Rights Management (DRM)	84	1932813403	$19.99	2006
BK7781351	Introduction to IP Audio	64	1932813519	$19.99	2006
BK7781335	Introduction to IP Television	104	1932813357	$19.99	2006
BK7781341	Introduction to IP Video	88	1932813411	$19.99	2006
BK7781352	Introduction to Mobile Video	68	1932813527	$19.99	2006
BK7781353	Introduction to MPEG	72	1932813535	$19.99	2006
BK7781342	Introduction to Premises Distribution Networks (PDN)	68	193281342X	$19.99	2006
BK7781357	IP Television Directory	154	1932813578	$89.99	2007
BK7781356	IPTV Basics	308	193281356X	$39.99	2007
BK7781389	IPTV Business Opportunities	232	1932813896	$24.99	2007
Legal and Regulatory					
BK7781378	Not so Patently Obvious	224	1932813780	$39.99	2006
BK7780533	Patent or Perish	220	0972805338	$39.95	2003
BK7769433	Practical Patent Strategies Used by Successful Companies	48	0974694339	$14.99	2003
BK7781332	Strategic Patent Planning for Software Companies	58	1932813322	$14.99	2004
Telecom					
BK7781316	Telecom Dictionary	744	1932813160	$39.99	2006
BK7781313	ATM Basics	156	1932813136	$29.99	2004
BK7781345	Introduction to Digital Subscriber Line (DSL)	72	1932813454	$14.99	2005
BK7727872	Introduction to Private Telephone Systems 2nd Edition	86	0974278726	$14.99	2005
BK7727876	Introduction to Public Switched Telephone 2nd Edition	54	0974278769	$14.99	2005
BK7781302	Introduction to SS7	138	1932813020	$19.99	2004
BK7781315	Introduction to Switching Systems	92	1932813152	$19.99	2007
BK7781314	Introduction to Telecom Signaling	88	1932813144	$19.99	2007
BK7727870	Introduction to Transmission Systems	52	097427870X	$14.99	2004
BK7780537	SS7 Basics, 3rd Edition	276	0972805370	$34.99	2003
BK7780535	Telecom Basics, 3rd Edition	354	0972805354	$29.99	2003
BK7780539	Telecom Systems	384	0972805397	$39.99	2006

For a complete list please visit
www.AlthosBooks.com

Althos Publishing Book List

Product ID	Title	# Pages	ISBN	Price	Copyright
Wireless					
BK7769431	Wireless Dictionary	670	0974694312	$39.99	2005
BK7769434	Introduction to 802.11 Wireless LAN (WLAN)	62	0974694347	$14.99	2004
BK7781374	Introduction to 802.16 WiMax	116	1932813748	$19.99	2006
BK7781307	Introduction to Analog Cellular	84	1932813071	$19.99	2006
BK7769435	Introduction to Bluetooth	60	0974694355	$14.99	2004
BK7781305	Introduction to Code Division Multiple Access (CDMA)	100	1932813055	$14.99	2004
BK7781308	Introduction to EVDO	84	193281308X	$14.99	2004
BK7781306	Introduction to GPRS and EDGE	98	1932813063	$14.99	2004
BK7781370	Introduction to Global Positioning System (GPS)	92	1932813705	$19.99	2007
BK7781304	Introduction to GSM	110	1932813047	$14.99	2004
BK7781391	Introduction to HSPDA	88	1932813918	$19.99	2007
BK7781390	Introduction to IP Multimedia Subsystem (IMS)	116	193281390X	$19.99	2006
BK7769439	Introduction to Mobile Data	62	0974694398	$14.99	2005
BK7769432	Introduction to Mobile Telephone Systems	48	0974694320	$10.99	2003
BK7769437	Introduction to Paging Systems	42	0974694371	$14.99	2004
BK7769436	Introduction to Private Land Mobile Radio	52	0974694363	$14.99	2004
BK7727878	Introduction to Satellite Systems	72	0974278785	$14.99	2005
BK7781312	Introduction to WCDMA	112	1932813128	$14.99	2004
BK7727879	Introduction to Wireless Systems, 2nd Edition	76	0974278793	$19.99	2006
BK7781337	Mobile Systems	468	1932813373	$39.99	2007
BK7780534	Wireless Systems	536	0972805346	$34.99	2004
BK7781303	Wireless Technology Basics	50	1932813039	$12.99	2004
Optical					
BK7781365	Optical Dictionary	712	1932813659	$39.99	2007
BK7781386	Fiber Optic Basics	316	1932813861	$34.99	2006
BK7781329	Introduction to Optical Communication	132	1932813292	$14.99	2006
Marketing					
BK7781323	Web Marketing Dictionary	688	1932813233	$39.99	2007
BK7781318	Introduction to eMail Marketing	88	1932813187	$19.99	2007
BK7781322	Introduction to Internet AdWord Marketing	92	1932813225	$19.99	2007
BK7781320	Introduction to Internet Affiliate Marketing	88	1932813209	$19.99	2007
BK7781317	Introduction to Internet Marketing	104	1932813292	$19.99	2006
BK7781317	Introduction to Search Engine Optimization (SEO)	84	1932813179	$19.99	2007
Programming					
BK7781300	Introduction to xHTML:	58	1932813004	$14.99	2004
BK7727875	Wireless Markup Language (WML)	287	0974278750	$34.99	2003
Datacom					
BK7781331	Datacom Basics	324	1932813314	$39.99	2007
BK7781355	Introduction to Data Multicasting	104	1932813551	$19.99	
BK7727873	Introduction to Data Networks, 2nd Edition	64	0974278734	$19.99	2006
Cable Television					
BK7781371	Cable Television Dictionary	628	1932813713	$39.99	2007
BK7780536	Introduction to Cable Television, 2nd Edition	96	0972805362	$19.99	2006
BK7781380	Introduction to DOCSIS	104	1932813802	$19.99	2007
Business					
BK7781368	Career Coach	92	1932813683	$14.99	2006
BK7781359	How to Get Private Business Loans	56	1932813594	$14.99	2005
BK7781369	Sales Representative Agreements	96	1932813691	$19.99	2007
BK7781364	Efficient Selling	156	1932813640	$24.99	2007

**For a complete list please visit
www.AlthosBooks.com**

Printed in the United Kingdom
by Lightning Source UK Ltd.
135394UK00001BF/18/P